Collected Essays on Sound System Engineering

Volume I

Michael Lawrence

Contents

for Bills

(all three of you)

Introduction

In August of 2022, I released my book *Between the Lines: Concepts in Sound System Design and Alignment.*

My goal in writing that book was to provide a general foundational introduction to topic, and to familiarize readers with the concepts and practices that will empower them to be successful when designing, deploying and tuning sound systems, through the lens of my own views and experiences. It was important to me, in writing that book, to take a utilitarian, manufacturer-agnostic approach to the science and math that underpins this work; such that readers from all over the world, with access to so many varied makes and models of equipment, could benefit from the ideas and employ them in their own work as they saw fitting.

Since then, I have been posting a series of articles and essays on my professional website: technical explorations into specific points of curiosity, often aimed at readers of a higher level of technical knowledge and experience. As the amount of content on my website grew, I was often asked whether I would update *Between the Lines* to include the new information. I got so far as to finish writing a second edition of the book before realizing it didn't make sense: while *Between the Lines* is intended to serve as an "on-ramp" to the topic, it accomplishes that by being approachable both in terms of length, depth, and the technical complexity of the topics discussed within. By contrast, much of the material posted to my website was more niche, more specialized, relied on more advanced engineering concepts, or dealt with specific products.

Nonetheless, I continued to receive requests to publish that additional content for wider distribution. Eventually, I decided to leave *Between the Lines* alone, and to meet these requests by offering a curated collection of my more advanced articles and writings on topics within sound system design in a package as collected essays, following the form (although certainly not the intellectual prowess) of the audio engineering giants of generations past, including Richard Heyser and Harry Olson, whose collected works are an indispensable part of any complete audio engineering bookshelf.

In this book, you will find twenty essays I have published on my own website, precisionaudioservices.com, and some additionally within the pages of *Live Sound International* magazine, since the release of *Between the Lines*. They are generally shorter, more focused, and more technical in nature, with each having generally narrower applicability to a single topic or situation. As such, all of this material has been previously published in some form or another, and the content that is currently available for free on my website will continue to be available at no cost for the foreseeable future. If you are discovering this book without having read *Between the Lines*, you may want to give that a look and then come on back here, as the essays in this book build off the concepts and terminology discussed in that one.

Hopefully it goes without saying, but just in case — these essays are my personal opinion, based on my experiences and experiments. Your mileage may vary — and I could be completely wrong. Photos and screenshots of many different products by many different manufacturers appear within the following pages, and all of it is intended to serve as an example of a real-world use condition that I encountered for educational purposes, not an endorsement or anything of the sort.

Some people just enjoy having things in book form. Not to worry — I am one of them.

As my investigations continue, perhaps we will someday find ourselves here again, discussing Volume 2.

Michael Lawrence

Fall 2026

One

Beyond Coverage Angle

Why vertical coverage angle doesn't tell us the whole story

When choosing a loudspeaker or designing a loudspeaker array that will produce appropriate coverage for a given audience geometry, a few key parameters guide the way. The first — and most widely recognized — is vertical coverage angle. This determination is simple — from the loudspeaker's position and perspective, what is the angle inscribed by the coverage area? In other words, how many degrees of coverage do we need to get from the most distant listener to the closest?

Figure 1-1 shows a generic arena side profile ("section view"). We can see that the amount of vertical coverage required depends on which seating sections are to be sold for the performance.

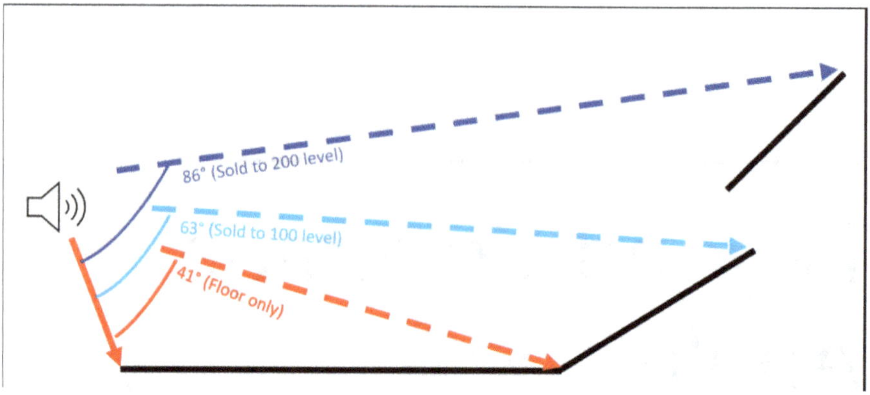

Figure 1-1

Now it might seem that as long as we have the proper vertical coverage angle, we're all set for uniform coverage. Not so fast! Knowing how much total coverage we need is a good start, but that's not all the information we need for uniformity.

Range Ratio

The next parameter to consider is *Range Ratio*. Range ratio describes the distance from the source to the furthest seat, compared to the distance from the source to the nearest seat. This helps us determine how asymmetrical we need our coverage to be. A low range ratio (like 2:1) means that the furthest listener is only twice as far away as the nearest. We need twice as much SPL (6 dB) to go towards the back if we want to create uniformity. With a larger range ratio like 6:1 (15 dB) we need to design an array that has a *lot* more output towards the top than at the bottom in order to achieve uniformity from front to back.

Figure 1-2 shows the geometry of three difference audience areas with range ratios from 2:1 (6 dB) to 11:1 (21 dB). Despite the fact that all three have the ***same exact vertical coverage angle***, they have drastically different needs for SPL distribution if our goal is to achieve coverage uniformity — the deepest audience area would require an array that has 15 dB more output towards the top than the shallowest audience area.

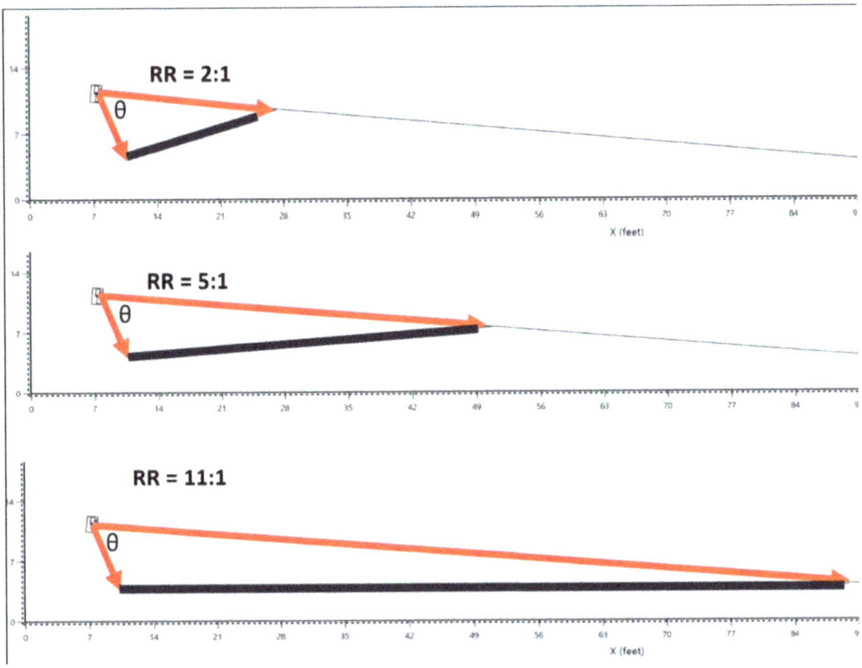

Figure 1-2

In order to create asymmetrical dispersion of high frequencies within a given coverage angle, we can manipulate the splay angles between the elements in an array. Smaller inter-element splays mean more elements radiating into the same angular space, and thus more energy headed in that direction. For example, we could achieve 12° of coverage with a single 12° source, two elements at 6°, three elements at 4°, four elements at 3°, six elements at 2°, or 12 elements at 1°. In all cases, we will be able to supply the required 12 degrees of vertical coverage, but SPL increases as the splays decrease and the element density increases.

Setting Splay Angles

Now we can tailor the allocation of elements within the array as necessary: the more distant sections of audience geometry will be covered by more elements at smaller splays, while the closer audience sections can be covered by fewer elements at larger splays. As a geometric shorthand for this, **we can simply adjust the splay angles until the center lines impact the audience area at equally**

spaced intervals.

This will get us to a point where we should need only minimal electronic processing (high frequency shading) to achieve uniformity from front to back.

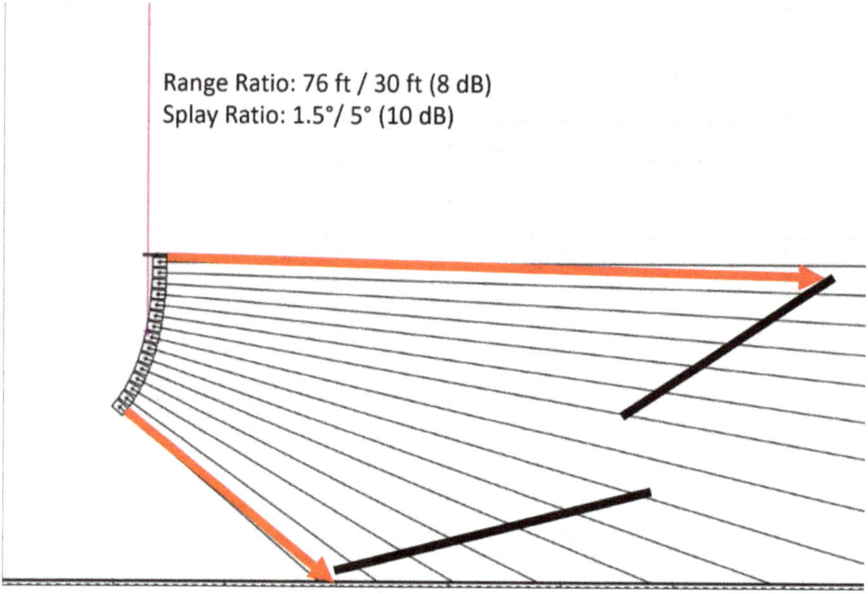

Figure 1-3

In Figure 1-3, splay angles have been adjusted to create evenly spaced centerline impacts from back to front (this is how auto-splay algorithms work). This dictates the inter-element splays and overall array curvature, with smaller splays towards the top and wider splays towards the bottom. Since the range ratio is relatively gentle (about 2.5:1, or 8 dB), so is the resulting curvature. Figure 1-4 shows us that we have indeed created quite uniform HF coverage (4 kHz, 1 octave). The prediction is colored in 3 dB increments, and indicates that all the listeners are now within +/- 1.5 dB at 4 kHz.

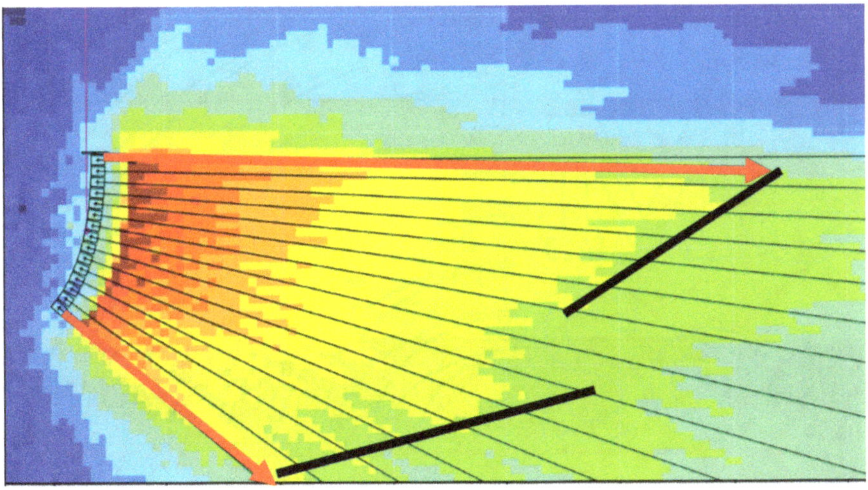

Figure 1-4

Going Long

Now let's push the stage further back into the venue so we can sell more seats. The vertical coverage needs have not changed, but the range ratio has been drastically increased. If we use the same splays / curvature, our energy distribution is now a poor match for the audience geometry, with a 4 color (12 dB) drop from front to back. (Figure 1-5).

Figure 1-5

Let's readjust the splay angles — closing up towards the top, and opening up more towards the bottom, and redistribute the splay within the same total vertical coverage. Now our splays run from 1° to 10° (20 dB) which is a ratio of 1:10 compared to the 3.3 we had with the shallower audience area. Figure 1-6 shows we're headed in the right direction again.

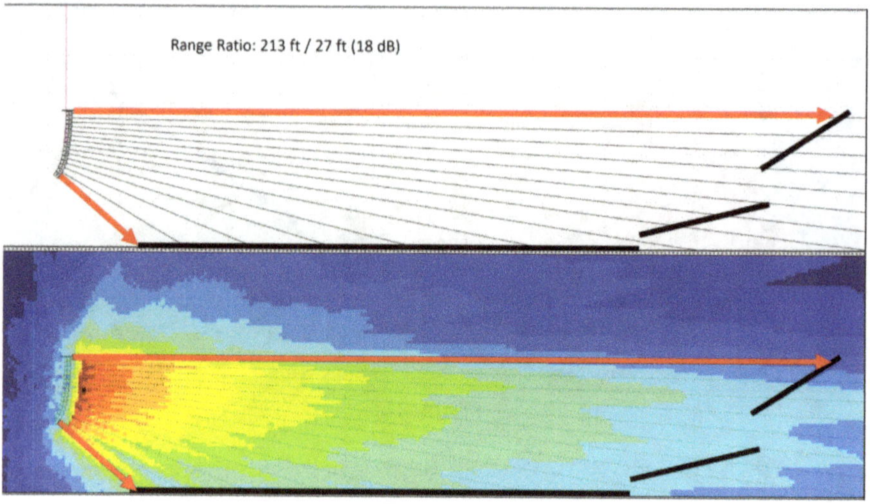

Figure 1-6

Overshoot

Now we have a new problem: our narrower splays at the top of the array are putting a harder edge on the HF coverage, and we're missing the last few rows.

We need to tip the entire array up a few degrees and "overshoot" a few boxes to keep them within the coverage. Tipping up by 3 degrees puts our third box on axis with the last row of seats (2 boxes of overshoot) and brings those seats back into coverage.

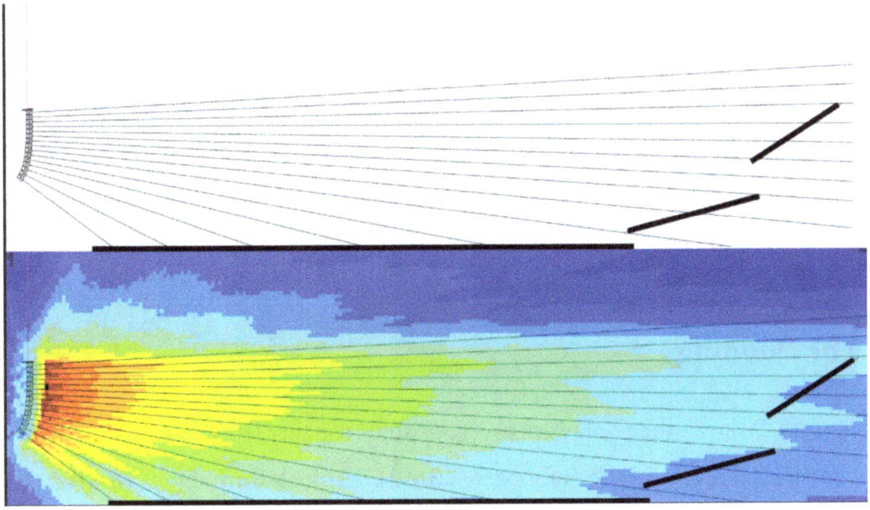

Figure 1-7

Now we are mechanically optimized for HF uniformity and a small amount of HF shading on the bottom few boxes will bring us over the finish line.

Line Length and LF

We have one final consideration: so far we have been looking exclusively at HF behavior, where the splay angles are the majority shareholder in coverage behavior. Our coverage at low frequencies is determined by line length. At 250 Hz, we see this array length is a reasonably good fit for the coverage shape we are trying to create (Figure 1-8).

Figure 1-8

If it were too wide, we could narrow it by making the line longer (more boxes), which would also mean closing up splay angles to keep the proper overall vertical coverage angle at HF. If it were too narrow, we could shorten the line by taking a few boxes off, and open up splays to restore the proper HF vertical coverage.

In this case we have very good uniformity (+/- 1.5 dB) except for the 200 section which is 3 dB down. If our goal is to preserve tonality (minimize spectral variance) we can easily adjust the HF shading on the top few elements to maintain the overall tonality at a lower level.

The Pudding

Figure 1-9 shows the final results — consistent tonality from front to back with a controlled level variance (about +/- 3 dB) despite a formidable 18 dB range ratio. This is accomplished predominately mechanically, by properly specifying total vertical coverage, interelement angles / curvature and line length. The only processing used was HF shading (shelving filter) on the top 4 and bottom 3 elements.

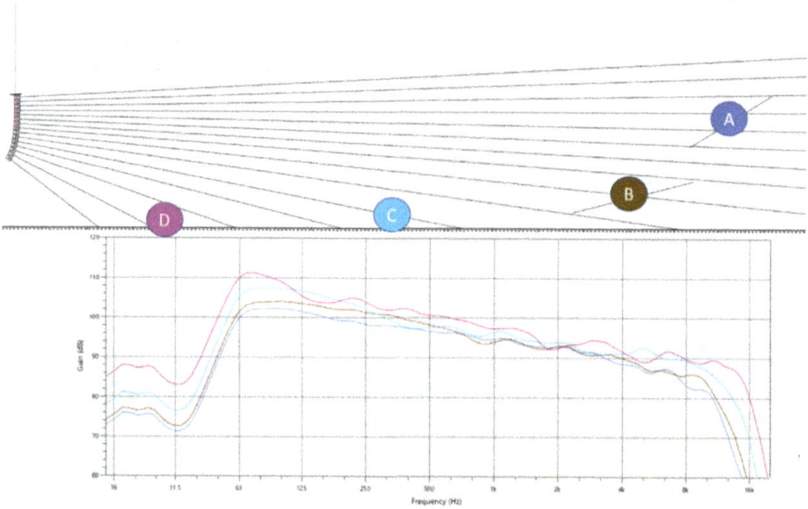

Figure 1-9

Design Parameters

To achieve optimal coverage, line length, overall vertical splay, and inter-element splays / curvature must be able to be manipulated independently. **Simply deploying an array with the proper vertical coverage angle does not mean it is an appropriate fit to create uniformity in the audience geometry.** We must also consider range ratio (and therefore splay ratio) and line length. Systems which inexorably tie two or more of those design parameters together inherently create design compromises.

Constant curvature systems, for example, bind vertical coverage, interelement splay and line length all together. If the audience plane has a low range ratio — that is, all the listeners are about the same distance from the source — a constant curvature may give satisfactory result at HF. However, the total required vertical coverage angle must then dictate the number of elements, since splay angle is fixed, which by extension means that we lose control of line length as a design parameter, and we must simply accept the LF coverage shape as-is. There is no way to improve LF directivity by increasing line length without increasing vertical coverage angle at the same time, resulting in improper coverage for the application.

Two

Steer Me Right: Understanding Beamsteering

The conceptual foundations of modern beamsteering technology

This essay is the first in a two-part series that examines basic beamsteering in the context of modern sound systems.

Part one (you are here) will explain the fundamental concept of using time delay and phase offset between array elements to affect directivity, and the associated effects in the time domain.

Part two will then explore how these principles are employed by three major loudspeaker manufacturers in under-the-hood processing to achieve the desired result, and look at the practical pros and cons of each approach, and of electronic beamsteering as a whole.

Meet the Players

To illustrate the fundamental mechanism at play, let's look at a line of subwoofers spaced by 3 feet (between acoustic centers, not between cabinets). Let's view their behavior at 63 Hz:

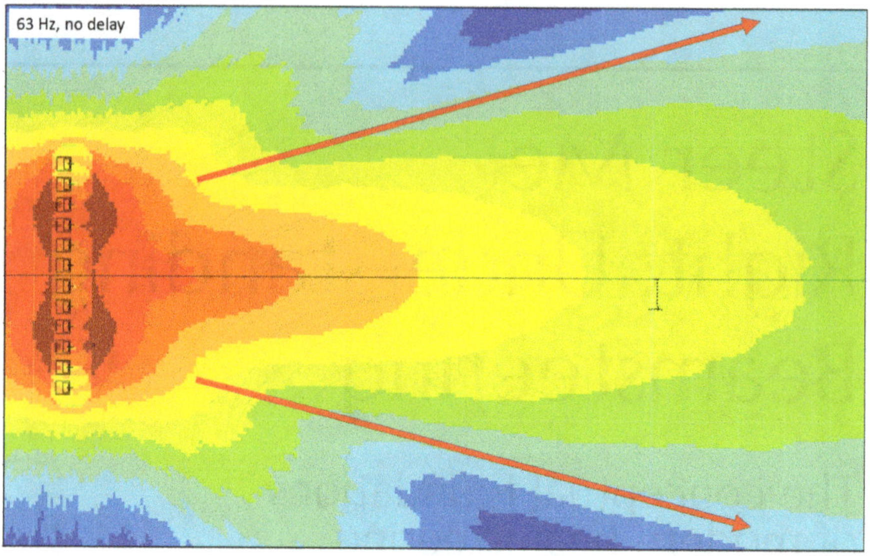

Figure 2-1

This is a coverage shape that mixers affectionately refer to as Power Alley. In front, the arrivals from all 12 elements are very close in time, and so sum together quite well. Once we get into the "blue rivers" off to either side, we have moved far enough away from center that we now have significant time offset between arrivals, and the staggered arrival times start to create a cancellation. I have added red arrows to indicate the approximate coverage angle of this array.

The astute observer will notice a "gull wing" side lobe forming even further off axis — over here, the time offset between element arrivals has increased even further, to the point that the more distant boxes are starting to end up a full cycle behind the closer ones — they are getting "lapped" — and so starting to sum again (they are in phase, but not in time). Considering that a full cycle at 63 Hz is about 16 ms, it makes sense given the length of the line that we would have the late arrivals showing up 16 ms behind the early ones.

It's also important to extrapolate based on this information that we get a different pattern at different frequencies. For example, if we view the same prediction at 125 Hz (an octave up):

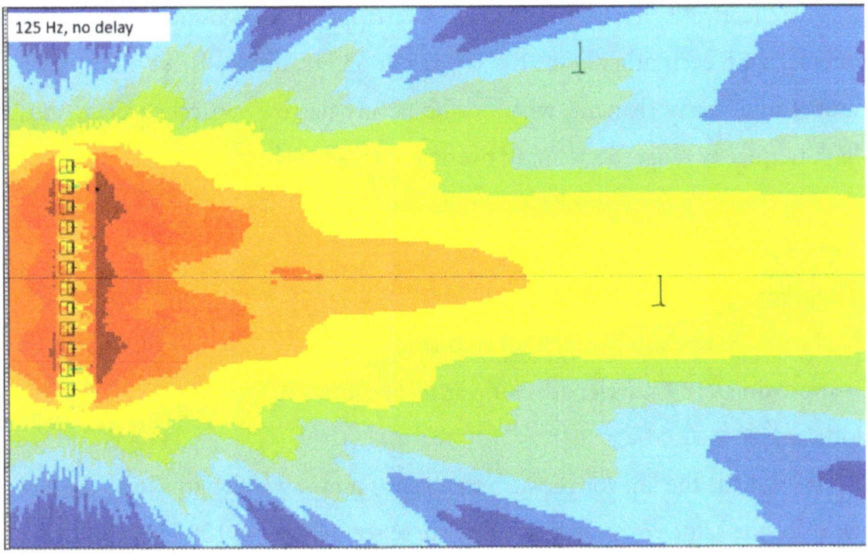

Figure 2-2

The main coverage angle has narrowed by about half. This makes sense — 125 Hz has a period of 8ms, which is half of the period of 63 Hz, and so as we start to move off axis, we only need to go half as far to accumulate enough time offset for our summation to break down.

This idea ends up becoming one of the most important concepts in the discussion to come: **a given amount of time offset (delay) will cause different amounts of phase offset at different frequencies.**

Given the narrow bandwidth of subwoofers (practically speaking, about 2.5 octaves) we have a little more leeway when using delay to steer the coverage, but this is a luxury we will not be afforded when we move to full range sources (8 octaves). The "right" amount of delay to accomplish our desired steering at a certain frequency will be ineffective lower in frequency, and create disastrous amounts of phase offset a few octaves up.

Another concept that is important to note before we move forward: a single subwoofer is, more or less, omnidirectional at 63 Hz. When we start arraying them together, their coverage patterns all overlap significantly, and it's this overlap that creates the beam narrowing (power alley). Manipulating the coverage pattern electronically, i.e. by playing with timing / phase offsets between elements — is only effective if those elements have overlapped coverage such that they can interact with each other. This is a safe bet at 63 Hz, but less so at 4 kHz with a modern line array element, which exhibits very narrow vertical coverage in the high frequency range. So we need overlap to make this work.

Another way of thinking about this: within the LF range where the elements are highly overlapped, line length is what determines the resulting coverage angle. The reason the coverage is half as wide an octave up is that the line length is doubled when viewed as a function of frequency. In other words, our subwoofer line is approximately 2 wavelengths long at 63 Hz, and approximately 4 wavelengths long at 125.

If we made the line longer such that it was now 4 wavelengths long at 63 Hz, we'd see the same coverage angle that we're seeing now at 125 Hz (Figure 2-2), and 32 Hz would be making the same coverage angle that 63 is in Figure 2-1.

Going forward, successful beam steering with full-range sources will rely upon finding a mechanism that allows us to affect the time / phase offsets between sources in the frequency range we want to steer, without disturbing the timing / phase relationships at other frequencies. In other words we will need frequency-dependent processing of some sort.

Beam Spread

Conceptually, we have a situation where our sources are effectively omnidirectional, so they all have a *similar* level regardless of where we choose to stand. The difference in arrival time is the major player in determining the combined coverage shape. Without any delay in the sub array, all the arrivals are in time out in front ('on axis') and time offset (and therefore phase offset) accrue as we move radially to the side.

Grab a laser disto, stand out in front of the array and measure the distance from your position to each element in the array. The numbers will all be similar. Now stand 90° off to the side (at the end of the line) and repeat the experiment

— a much larger variance in distance, which means much more time offset.

Geometrically speaking, the time offset is *least* out in front, which is why we have the most summation there (power alley), and it is *most* on the sides of the array, where the cancellation is strongest.

Simply: least time offset in front, most time offset on the sides.

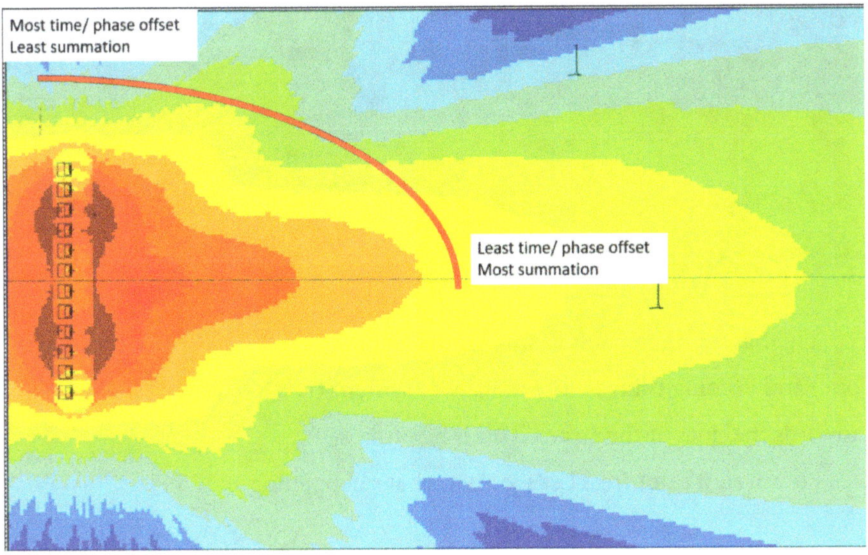

Figure 2-3

We can disrupt this interaction a bit by adding progressively more delay to elements that are further from the center of the array. Now they are not all arriving quite so close in time out in front — we have staggered the arrivals a little bit more, which should break down the summation a bit. Similarly, as we move off axis, the elements on the ends of the line won't be *quite* as early to the party as they were before — because we delayed them. So by adding a delay taper towards the ends of the array, we increase time offset in the front, and decrease time offset on the sides.

Let's try this with a 1ms linear taper. We'll leave the center pair at 0 ms, add 1 ms of delay to the next pair, 2 to the next, and so forth, until the outermost pair has 5 ms.

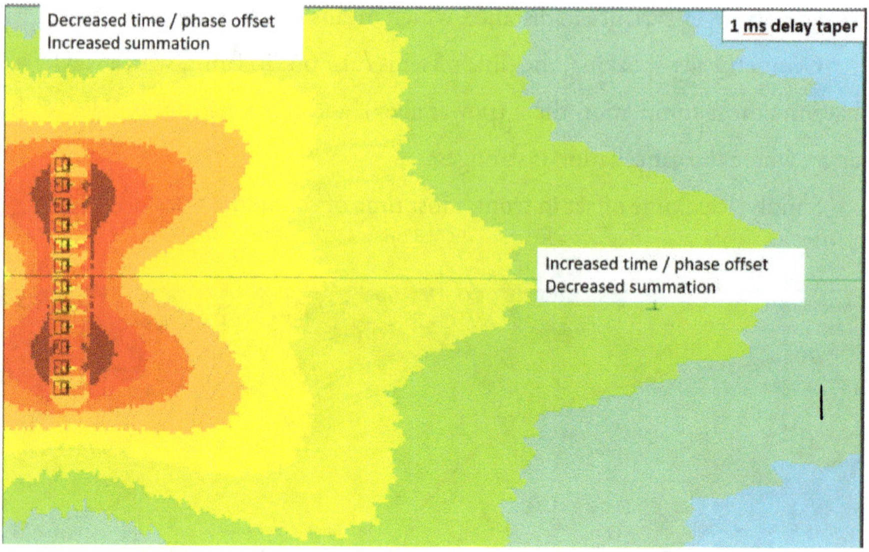

Figure 2-4

This has the desired result: widening out the coverage angle (compare to figure one) and keeping things more radially consistent (less level drop as we move off towards the sides of the array). This is helpful.

However, it's not free. Let's look at what's happening at 125 Hz:

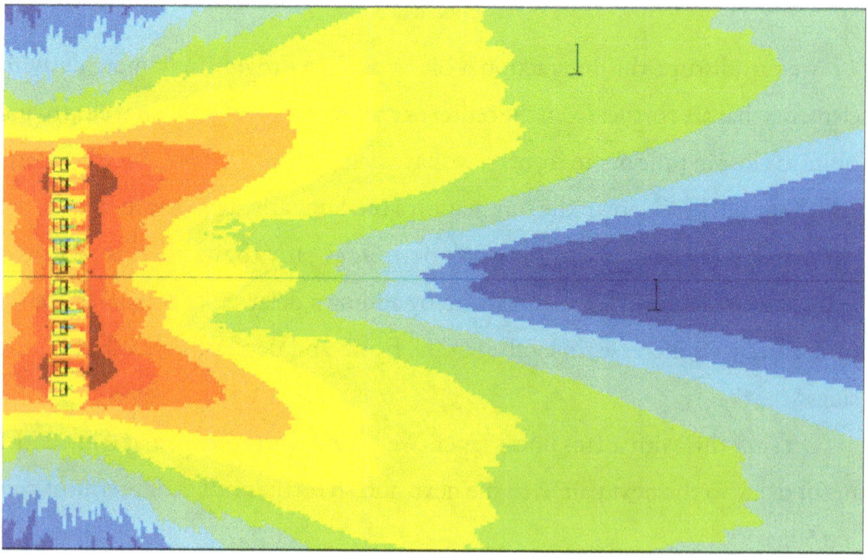

Figure 2-5

We have too much of a good thing. The time offsets that were helpful at 63 Hz are too much at 125 Hz, because they create twice as much phase offset out in front, resulting in a cancellation going forward. If we want to find a way to make this steering work with full range systems, we will need to find a way to adjust time / phase offset per frequency.

Impulse Response

Before we move on, let's take a second to examine the time domain ramifications of what we're doing. By staggering the arrival times of elements out front, we have energy from the sub array arriving over a longer period of time at mix position.

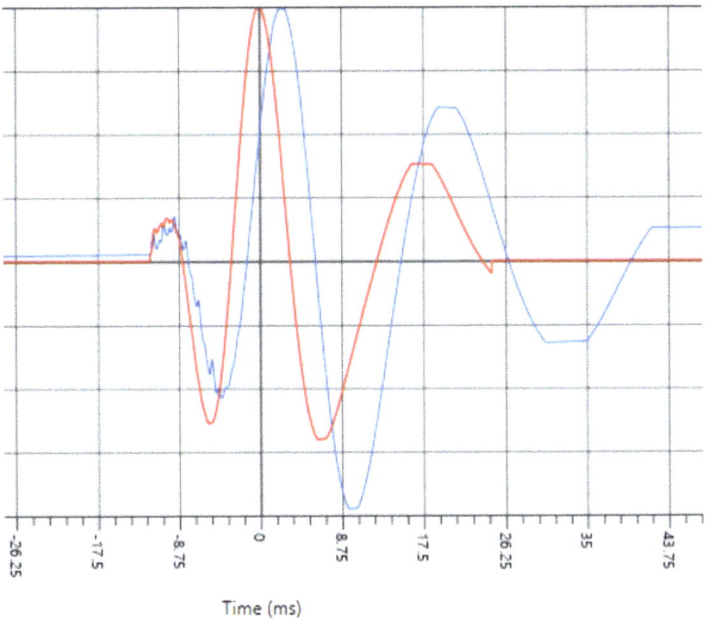

Figure 2-6

The peak in the IR is a bit later in time for the delay-tapered array (blue) compared to the non-steered array (red), which makes sense, as more of the energy is arriving later, and also the overall impulse response is lengthened by more than 15 milliseconds. This can be referred to as 'time smearing' and is sometimes described by FOH mix engineers as "pillowy" or "squishy", which seem to be subjective descriptors of the fact that the impulse response of the subwoofer array is longer, and less "sharp" or "tight."

As a practical note, gathering clear impulse response data at these low frequencies is extremely challenging in a real-world acoustic environment, a topic that we'll be revisiting later in this series. Fortunately, the MAPP XT prediction is helpful in this regard. (Note: As of writing, the other two prediction software platforms we will be using for this series do not offer the ability to view impulse response, although I have submitted feature requests for such.)

By the way — what do you think is happening to the IR of the sub array over off to the sides, where the blue rivers were in Figure 2-1? The opposite — the blue river was caused by staggered arrivals over time, and our delay tapering has made them "less staggered". The blue rivers filling in mean less time-domain stagger and a tightening of the IR.

Steer Up

Now let's try a variation on the same thing. Instead of applying delay from the center outward, let's start one end at 0 ms and apply progressively more delay in 0.5 ms steps as we move to the other end, ending up at 5.5 ms on the last element.

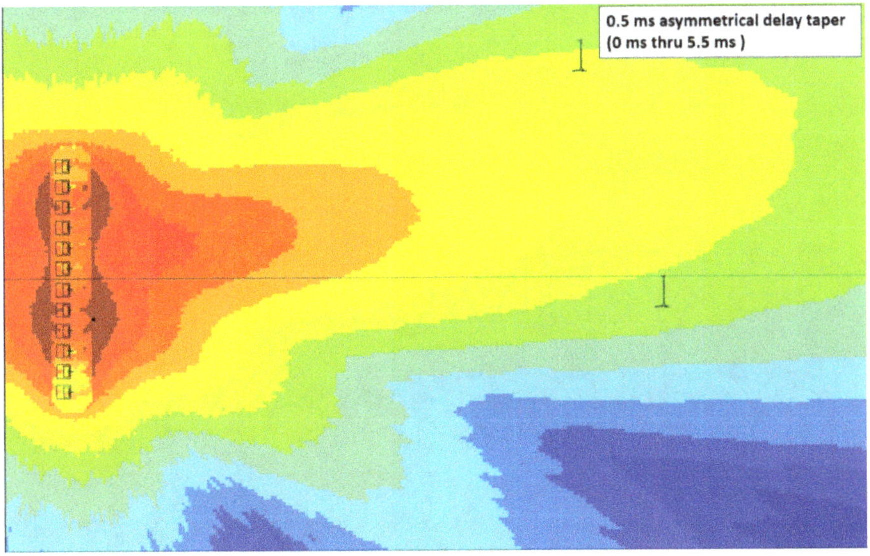

Figure 2-7

The element closest to the bottom of the image has no delay, and the element closest to the top has the most. Remember what caused power alley — equal-level, equal time. We haven't changed levels, but we have changed where "equal time" is in the space, and moved it closer to the top of the image by giving the boxes lower down a head start. In effect, we have pointed power alley in a different direction.

Remember that the frequency-dependent effects still hold true: coverage will still be wider at 32 Hz and narrower at 125 Hs, and that's a limitation of relative line length and using frequency-independent delay.

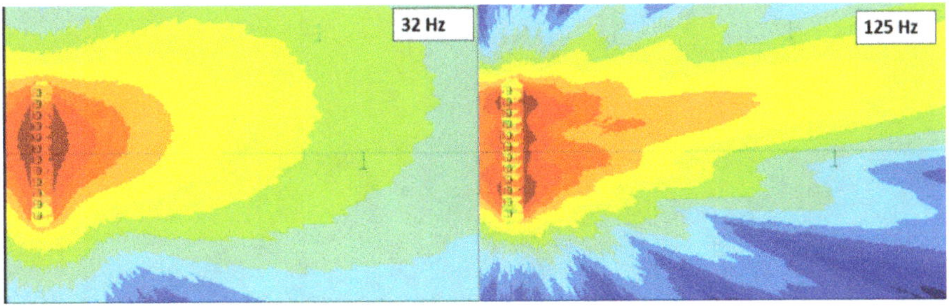

Figure 2-8

Now that we've look at the basic concepts, in the next essay we'll learn how manufacturers are putting them to work in beamsteering algorithms.

Three

Steer Me Right: Principles In Practice

Example manufacturer implementations of beamsteering technology

In the previous installment, we learned the basic concepts of how to apply delay to elements in an array to steer the coverage, and studied the effects of that steering on the impulse response, as smearing caused by spreading the arrival times. Let's begin this exploration by understanding the problem we're trying to solve.

In FIGURE 3-1 below, we can see an example of an asymmetrical coverage challenge, where we need more output towards the top of the array to reach the farther listeners. This is easy to accomplish in the high frequency range by using the appropriate splay angles (see Beyond Coverage Angle for more) but as the figure confirms, the low-mid coverage will always emanate from the geometric center of the array. In many common geometries, this is not a pressing issue, but in a situation such as this, where we need rather asymmetrical coverage, we could achieve higher uniformity if we had a way to steer the low-mid lobe upward so it was a better match for the HF. In other audience architectures, it's possible that the low-mid lobe may end up pointed in the right direction, but too severely narrowed to match the HF shape well.

In other geometries, a combination of both problems may manifest. In all cases, the issue is the same in general terms: the low-mid coverage shape of the array is not a good match for the HF coverage shape. While the short wavelengths at high frequencies respond well to splay angles and the directivity of the horns and waveguides (magnitude steering), the array elements are highly overlapped at LF and so "steer by committee" (phase steering). Anything we do to try to rectify this issue in the low mid must therefore rely on the interaction between the elements to create the desired beam shape.

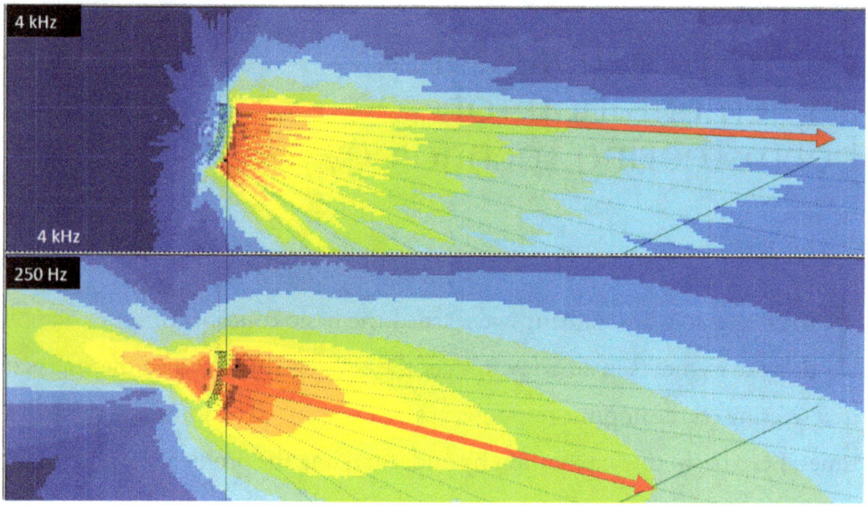

Figure 3-1

Of course, we could try to do this mechanically by tipping the entire array upwards, however that would result in all the HF going in the wrong direction, so that's not a good tradeoff.

We can apply what we learned in Part 1 by applying delay incrementally up the array, with more on the top boxes, to try to steer that low-mid lobe upward. We used a handful of milliseconds to do this steering in the sub range. Two octaves (4x the frequency) above that at 250 Hz, we'll have to use a quarter of the amount of the delay time to get a comparable effect. So let's try a 0.25 ms taper per box, which means box 12 at the bottom of the array starts at 0 ms, and box 1 at the top ends up with 2.75 ms.

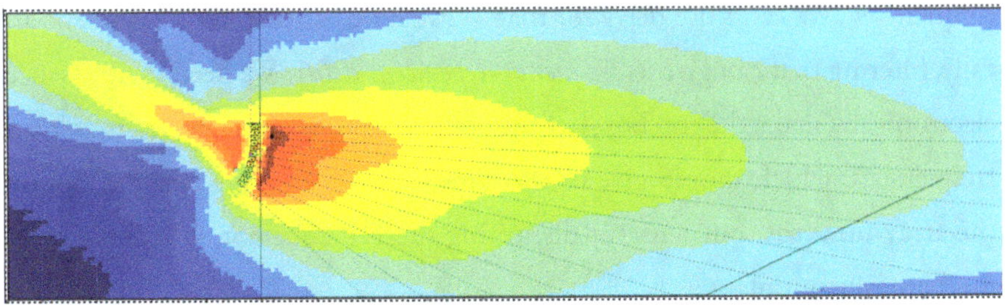

Figure 3-2

This actually works brilliantly! The coverage shape at 250 Hz is now an excellent match for the 4 kHz shape. Only one small problem — it's completely destroyed the behavior of our system higher up in the spectrum. That 0.25 ms delay offset is only an eighth of a cycle (45°) offset at 250 Hz, which is why it works so well for steering, but it's causing mayhem in the higher frequencies — for example, we now have adjacent array cabinets firing a full cycle apart at 4 kHz. That delay time creates a different phase offset at different frequencies, completely disrupting the summation behavior and steering energy off in weird directions. It also — and this is kind of important — sounds terrible, because we've effectively built comb filtering into our sound system.

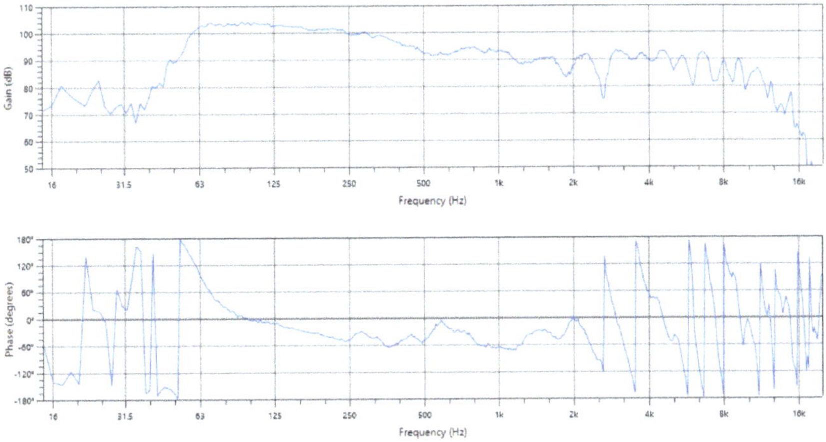

Figure 3-3

So what we need is a way to apply phase offset per element in the low-mid region to accomplish the steering, but without disrupting the response at higher frequencies. We can't use delay for that — we have to use filtering, and specifically, filtering that can give us *frequency-dependent delay*. We need to, in essence, electronically make the top boxes of the array further from the listeners at lower frequencies, which should give us the same coverage steering effect as tipping the array upwards, but without disrupting the HF behavior, which is already mechanically optimal.

About This Investigation

The remainder of this article is an examination of the steering tech used by three leading loudspeaker manufacturers: Meyer Sound, d&b Audiotechnik, and L-Acoustics. I have attempted to reference 'explainer' resources from all three manufacturers we'll be examining here — I think it's important to understand how the manufacturer has explained their toolset before doing our own evaluations of it.

An important note: everything discussed here is based on information made publicly available by the manufacturers and what is indicated in the measurement data presented below, and similar data can be produced by anyone with access to the equipment. The explanations and conclusions presented here are simply my educated guesses based on measurements and established engineering principles, and may be incomplete or incorrect — I have not requested review or comments on this essay from any of the manufacturers. I feel that it is part of my professional responsibility as a systems engineer to understand the tools we use in our work, and this investigation is my own exploration into that understanding. Readers are advised to draw their own conclusions based on the available information, and to consult the manufacturers in question for guidance on implementing these tools. Opinions presented are my own.

In addition, a practical caution is in order: directly measuring audio amplifiers involves high voltages that can be potentially harmful to both yourself and your audio interface — if you choose to undertake a similar study, proceed with care!

Low Mid Beam Control

We start by looking at Meyer Sound's low-mid beam-steering solution, termed Low Mid Beam Control (LMBC). The manufacturer's website hosts a brief "explainer" video that recaps the issues we've examined so far, as well as introduces Meyer Sound's solution in the form of LMBC.

LMBC uses staggered all-pass filters per element to create phase offsets in the low-mid region without disrupting the higher frequencies. Important note: the all-pass filter itself is not what causes the steering — any processing that is applied equally across the entire array doesn't change directivity. The mechanism here is that the all pass filter is tuned to a *slightly different frequency* per element, and that small phase offset between elements in the low mid region is what steers the coverage. As we move forward, keep in mind that in all cases, the absolute phase response of the array, or the loudspeakers within it, is not of interest here — what makes the steering work is the *difference in phase response from element to element*, so that's what we should focus on. By measuring each channel of the GALAXY loudspeaker processor directly, we can study the filtering.

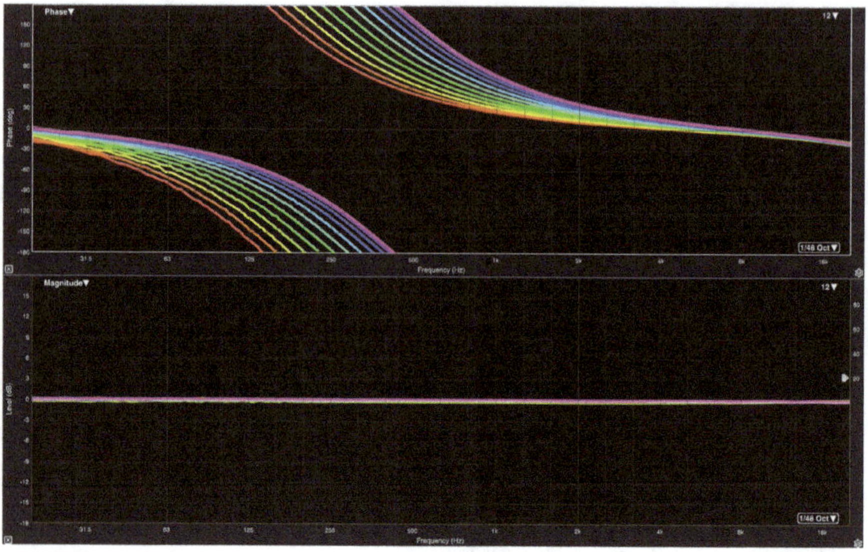

Figure 3-4

True to its name, the All Pass Filter has no effect on the magnitude trace. By rolling the phase trace 180° we can gain a more intuitive look at the staggered filtering. (Phase data is cyclical but a phase plot is a 2D graph so graphing it adds an edge *somewhere*, just like the difference between a globe and a map — there are no dragons off the edge of the map — it is seamless data — so too with the phase trace).

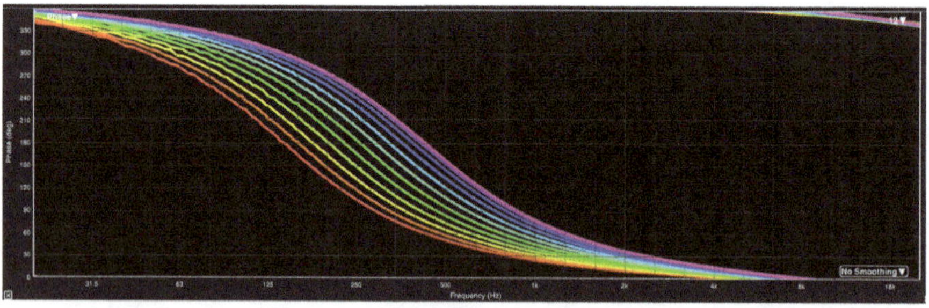

Figure 3-5

Here, the filtering creates about 17° of phase offset between elements at 250 Hz (equivalent to a delay taper of about 0.2ms) but the traces converge above and below that, causing minimal disruption at HF. Note that filter spacing will vary based on the LMBC parameters, but the mechanism is clear. What we have is frequency-dependent delay between elements in the low mid range, which is responsible for the steering. Setting the mode to "Beam Spread" would taper from the middle of the array outwards in both directions, instead of a taper from top to bottom as we see here. Keep this in mind — we'll see it pop up soon.

The biggest benefit of LMBC in my opinion is that the controls and the filter parameters are viewable for evaluation and adjustment by the systems engineer. That is, you can see exactly what it's doing, because the APF are displayed in the *Compass* control software, and a user can adjust the LMBC parameters while watching the filters move around, and can modify the behavior as they see fit. Having the filter parameters displayed also means it's straightforward to apply matching filtering to supporting systems (say, front fills) and opens the door for "out of the box" use cases.

A few examples: since the LMBC dialog displays the "average" APF parameters, I sometimes tweak the LMBC parameters for side hangs until the filtering lands on the same center frequencies, thus ensuring that the main and side hangs have a closely matched phase response and the transition region is smoother. This is, like all things, a trade off, but I appreciate the flexibility to make this call on a case by case basis as I feel is best for the situation at hand.

I've also been able to successfully use LMBC to "steer up" on a main array that was driven as a hybrid: 2-box zones at the top of the array and 1-box zones towards the bottom. Since LMBC requires the user to specify 1 or 2 box zones per array, it can't natively support this configuration, but since it shows you where the filters should go, I was able to generate what the filter parameters *would* have been for 2-box zones, and simply add them manually as appropriate to generate the same steering effect on my hybrid zoned array.

It might be a benefit to some that LMBC uses all pass filters, which invoke no additional delay into the transmission chain, unlike the FIR filters we're about to see in other implementations. However, I find this to be less of a concern in the context of large-format PA systems, where the propagation through the air is the majority shareholder in transmission time, unlike monitoring situations where low latency is critical.

The biggest drawback of LMBC in my opinion is that it's inherently limited: "steer up" is limited to array splays of 45 degrees or less, although there are ways to cheat that when desired, and although the two options are "Steer Up" or "Beam Spread," sometimes neither is a complete solution. It all depends on the geometry you're trying to cover, and in general, LMBC ends up being a more useful tool in sheds, amphitheaters and outdoor stages, where trim heights are relatively low and curvature is generally less, and I find it less useful in arenas where trim is higher and curvature is generally more extreme (it is not uncommon for a main array in an arena environment to have 75° of splay or more).

Take a closer look at the prediction above, and notice that the *rear lobe is steered up as well.* Beamsteering is like opening or closing an umbrella, and it will steer your side and rear lobes along with the main coverage lobe. We will revisit this point later.

Some people may also find the phase wrap introduced by the all-pass filter to be a drawback. If having a phase-linear loudspeaker system is important to you, then you probably don't want to sully that by adding a wrap. Remember that the phase response is related to the time domain response (impulse response) of the system, and that larger portions of the spectrum having a flat phase response mean that more of the frequency range is arriving all at once, and all of the products that use LMBC are phase-linear throughout most of their bandwidth as far as I know.

As we are about to learn, the FIR implementation used by other manufacturers avoids adding a phase wrap at the cost of some delay. However, the associated loudspeakers are not phase linear to begin with. In my opinion this is "six of one, half dozen of the other." In my personal opinion, the IR smearing as a result of the different per-box responses that we get when applying any sort of beamsteering is a far more audible issue than whether the loudspeaker in question has a phase wrap, but more on that later.

ArrayProcessing

Next on the list is the ArrayProcessing feature from d&b Audiotechnik. Here is the 'explainer' video — hopefully you will notice some common themes compared to the previous video — we are, after all, seeking to address the same issue in both cases.

According to the accompanying technical document, "ArrayProcessing creates individual sets of FIR and IIR filters for every single cabinet of the array, each of which requires a dedicated amplifier channel." The gist of it is that FIR filtering allows us to manipulate magnitude and phase independently, at the cost of delay. According to d&b's materials, ArrayProcessing adds 5.9 ms of delay to the signal chain — hardly a concern for most situations in my opinion.

To evaluate the mechanism, I created a similar venue and array architecture as the above example, and engaged a few ArrayProcessing slots, taking transfer function measurements of the amplifiers directly. I examined the effects on two different models of loudspeaker in two different array / venue configurations, and our focus here is on the general effects of enabling ArrayProcessing rather than evaluating its performance in each particular case. Evaluating the results is the purview of the system designer — our interest here is how it works. The following data shows the results of "Intrinsic" mode, which is the mildest form of ArrayProcessing in that it matches the natural front to back level drop of the array, rather than attempting to offset it electronically.

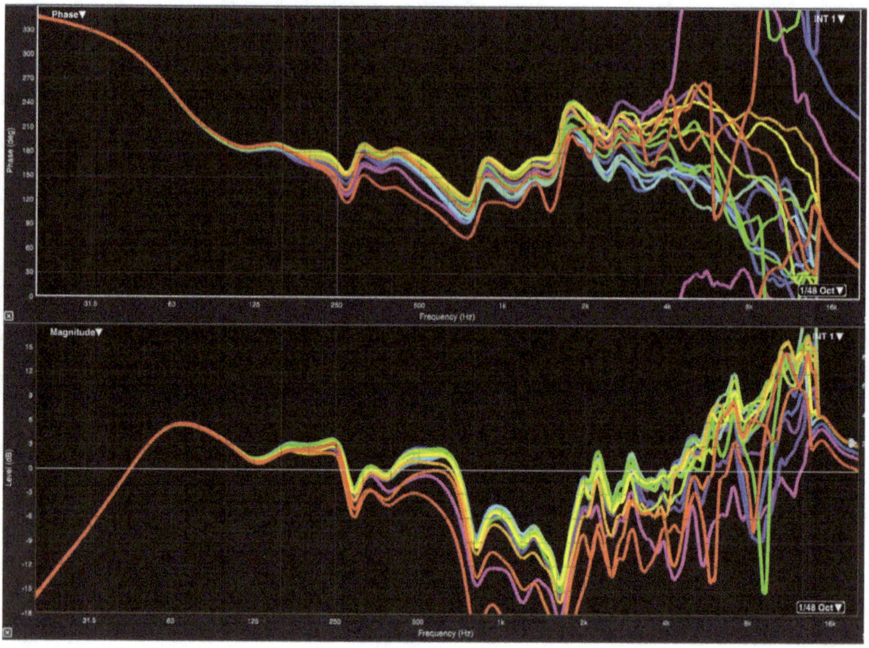

Figure 3 -6

There's a lot to look at here, but remember that since we're measuring through the amplifier / DSP directly, we are seeing not only the AP filters but also the "stock" manufacturer loudspeaker preset. We can eliminate the particular loudspeaker preset from the equation by exporting the "vanilla" (AP disabled) preset as a weighting curve, and then applying the *inverse* of that weighting curve to the magnitude traces of all these measurements. This will subtract out the response of the loudspeaker preset, and allow us to isolate and view *just the changes that AP filtering makes to the response of each element.*

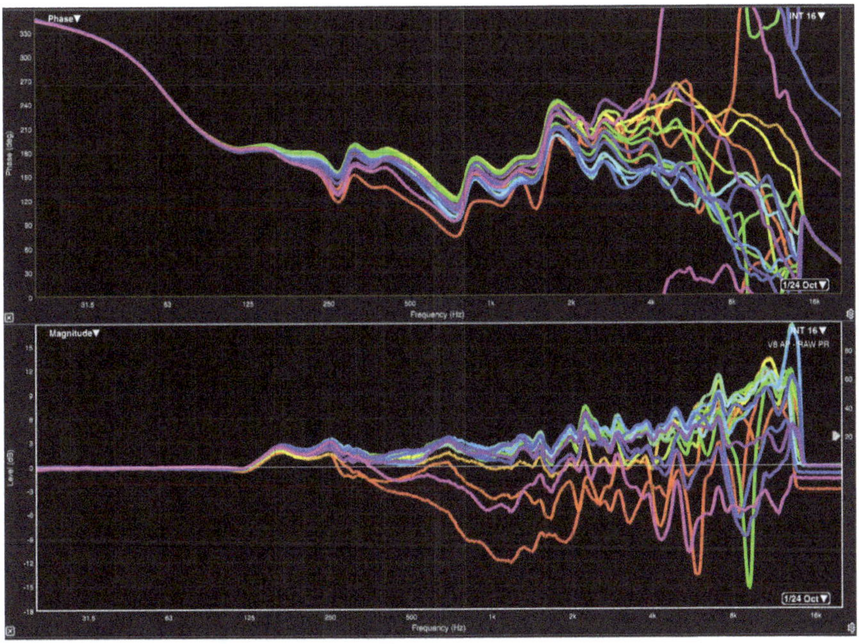

Figure 3-7

This is a little easier to parse — we can see that the processing is not limited to the low-mid band and instead extends up to approximately 13 kHz. As noted in the above-linked white paper, the processing includes air absorption effects and HF uniformity in its calculations which explains the HF filtering we're seeing. As an overall trend the processing has boosted HF overall, which would cause a perceptually brighter sound compared to it being disabled. Although the magnitude shading is interesting, our primary interest here is the phase offset between elements in the low-mid band.

Focusing our attention to the phase trace in the region between 250 Hz and 1 kHz, we see a familiar mechanism — a slight difference in phase between elements serving to widen and shape the directivity as the algorithm sees fit. This is accomplished without adding a phase wrap, and somewhat independently of the magnitude response — this is what has been bought by the addition of the 5.9 ms of delay for the FIR filters to do their thing. Interestingly the amount of phase offset between elements seems to remain more or less constant for a range of about 2 octaves, which sidesteps an inherent compromise of steering via all pass filter. There is also some magnitude offset in this range, but the phase offset is the majority shareholder of the steering force in the low-mid region of a line array. That being said, I did find the element-to-element response changes to be rather more severe than I would have expected given the filtering "constraints" described in the technical documentation.

It also occurs to me that this would be an excellent opportunity to phase-linearize the loudspeaker cabinet above 250 Hz or so — since FIR filtering is already in use, the necessary phase correction could be rolled into the same filtering without incurring any further delay, so I have to imagine there is some technical limitation in the implementation preventing this.

According to the technical documentation, the ArrayProcessing algorithm is "independent of array length, curvature, and system type. Any ArrayProcessing line array design will provide the same sonic characteristics." Although I do not have extensive firsthand experience evaluating this claim for numerous combinations of loudspeaker products, it is a welcome one because matching various system zones (mains, side hangs, front fills, etc) that are often comprised of different line lengths and different loudspeaker products typically requires some attention from the systems engineer.

ArrayProcessing can only be engaged when driving elements in single-box zones, which reduces the "pixelation" problem described at the conclusion of this article.

The ArrayProcessing dialog within ArrayCalc software includes a helpful *Result* tab showing the difference in magnitude response at impact points from front to back with and without the processing enabled.

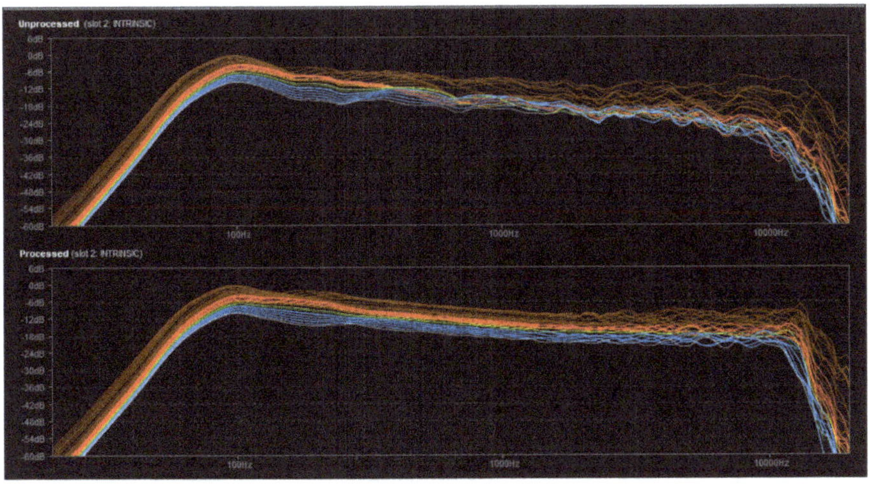

Figure 3-8

It seems that the algorithm only looks at the on-axis impact points (aka "landing strip") and doesn't seem to consider audience geometry impacts spanning the entire horizontal coverage of the horn, so some of the more severe deep cuts may be less than ideal in practice, and perhaps stricter "constraints" on the algorithm to further limit how far an element's response can deviate at any given frequency from that of its upstairs and downstairs neighbors would potentially increase coverage uniformity off axis and also further preserve headroom. Applying more drastic processing to try to reduce the front-to-back level variance of the designs I tested in turn produced more drastic filter responses, as one might expect.

Engaging ArrayProcessing also forces the user to implement air compensation, which is something I prefer to adjust manually on portions of the system rather than implementing a global change to the entire PA, because once the show starts I do not want any changes happening to the portion of the PA that the FOH engineer is mixing off of unless they directly request such a change — I just want to adjust the far-reaching elements of the system to keep in lockstep with what it sounds like at mix position.

It would be nice to have the option to restrict the filtering to the region below 1 kHz, to restrict the processing solely to low-mid beam steering, without having it affect the overall tonality of the PA or the high frequency shading. LMBC is inherently limited to the applicable frequency range, and the ArrayIntelligence

Optimization algorithm from Adamson offers the user an option to restrict processing bandwidth, whereas ArrayProcessing locks out per-amplifier EQ controls. (There is a somewhat time-consuming workaround for this.) As a general policy, I try to avoid using any functions that lock out the systems engineer from adjusting processing settings as they see fit. I consider this a limitation considering that the implementation doesn't allow the user to see or adjust the actual filtering being implemented into the system, just the predicted results.

Full Range AutoFilter

On to examining the Full Range Autofilter feature in L-Acoustics Soundvision. Some clarification is in order here: the AutoFilter in its early incarnation is simply an algorithm that attempts to increase HF uniformity by adjusting HF FIR filters per amplifier zone. The more recent iteration extends the filter effects to low frequencies. As of the time of writing, I was not able to find any written technical documentation from the manufacturer about the Full Range Autofilter, but I did find this webinar in which it is explained.

Seeing as engaging the full-range autofilter increases the delay through the system, and that it observably does not add a phase wrap to the response, we can surmise that it uses FIR filtering to stagger phase response per amplifier zone in the low-mid region. Since we are dealing with a three-way cabinet, it's a little harder to see in the data, because the steering region around 250 Hz spans the LF and MF "ways" of the amplifier processing, so we'll see a pair of traces this time. 6 traces for the 12-box array driven in 6 zones of 2 elements each. Remember — the exact model of loudspeaker here is not our interest — we are looking for how the response changes from zone to zone, which is the steering tool at work.

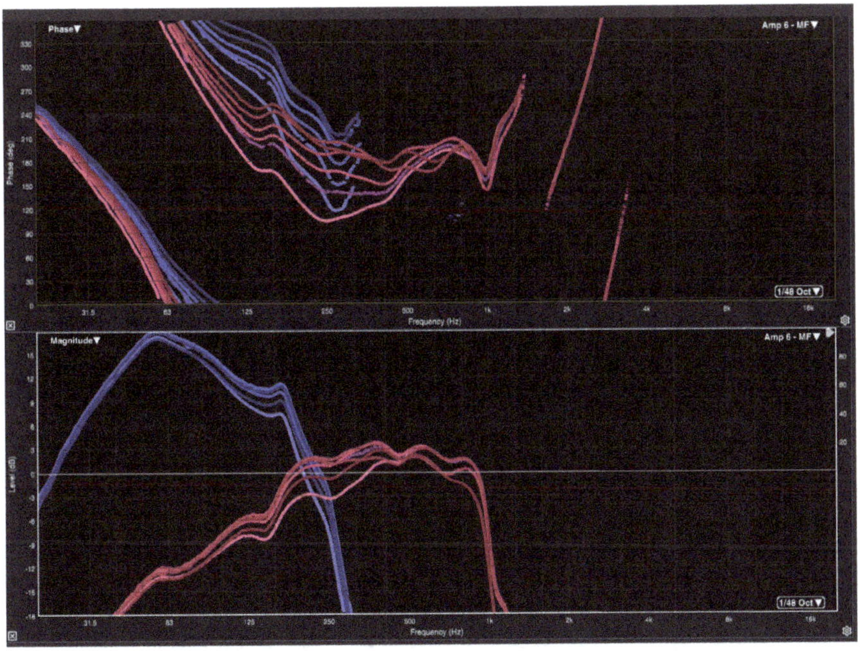

Figure 3-9

Again we see the familiar spreading of phase response per zone. Interestingly this filtering seems to be focused on a lower range than the previous tools we've examined — phase traces diverge between 80 Hz and about 500 Hz, and ride together below and above. The zone to zone spread is about 20 — 25° at 250 Hz, which is higher than we have observed up until this point, but this is likely owing to the fact that the processing is applied in 2-element zones in this case, and I suspect the "steps" would be even larger when applied to a circuit of three elements per amplifier. As we will discuss shortly, there are some drawbacks to using such physically large zones, as the acoustic "pixels" start to become the same scale as the wavelength they're attempting to steer.

One benefit to this particular implementation is that the HF FIR filtering (not measured for this exploration) is user-accessible and can be adjusted manually or disabled completely as desired without affecting the low frequency steering. This is a more ideal implementation in my opinion — limit the electronic steering to the frequency range over which it is required, and leave the HF shading and air compensation open to adjustment by the systems engineer if desired. This makes it much easier and more convenient to adjust to changing environmental conditions in certain portions of the system coverage, or respond to HF inconsistency — quickly and easily without disruption of coverage elsewhere.

However, I feel that the user-facing implementation is lacking, in that the current control is either on or off, with no way as far as I can tell to adjust the behavior of the steering beyond setting the coverage start and stop. This is surprising given the transparency of the HF portion of the autofilter, especially since the AutoFilter technology is described in the manufacturer's training slides as as "No 'black box' — all settings accessible / adjustable". And, as with above, no way to see the actual filtering being applied. My firsthand field experience with this tool is rather limited so I don't have much to say about it in terms of user experience.

Beamsteering — General Benefits and Drawbacks

In general, the benefit of these tools is obvious: it helps more of the energy go where we want it to go, and is a helpful tool to increase coverage uniformity in situations where mechanical solutions fall short. "Point the loud side toward the audience" is the principal tenet of sound system design, and these tools can help us do just that.

Inherent drawbacks are twofold — by "inherent" I mean that they are due to the mathematical realities of electronic beam steering, not a platform-specific drawback that could be resolved with a new software feature, etc. These are general considerations, not manufacturer-specific drawbacks.

Impulse Response Smearing

First: we pay a penalty in the time domain by implementing electronic steering, as we're manipulating the arrival time of energy on a per-element basis, and Part 1 of this series showed us how that smears the impulse response of the system. How severe or objectional this IR smearing is depends on a few factors — first, the preferences of the mix engineer. Some are bothered by it, some are not, as is true about most things in audio.

In a technical sense, the variables are overall array curvature (how much work we're asking the steering to do) and the "acoustic resolution" of the steering. IR consequences typically become more pronounced as overall array curvature increases, and as the zones get bigger. I find it less objectionable with less array curvature, single-box zones, and smaller-format loudspeakers (because the acoustic centers are closer together with respect to the wavelength being steered), whereas I find it more objectionable when the array has more curvature, is comprised of physically larger loudspeakers (15" rather than 12" cabinet height), and processing resolution is lower (2 box zones rather than 1).

This makes sense acoustically as well: single-box zones of a 12" array cabinet give us acoustic "pixels" that are less than 1/3 of a wavelength wide at 250 Hz. Two-box zones of a 15" cabinet are 3/4 of a wavelength wide. Think about how well a subwoofer array sums when the elements are spaced by a third of a wavelength, compared to when they are spaced by 3/4 of a wavelength, and you will see the inherent tradeoff here in side lobes and combing.

In addition to being audible, this IR smearing is measurable, but it is difficult since we are primarily talking about steering / smearing the low-mid frequencies, and the impulse response measurement mathematically gives far more emphasis to the high frequency range (3/4 of the FFT bins reflect the content of the top 2 octaves). This could be studied in the field with a high resolution IR measurement (large FFT, period-matched stimulus and lots of averaging) and then band-limiting the resulting IR to the low-mid region with a user-defined filter. A project for another day. In the meantime, I would like to see the inclusion of Impulse Response data in more manufacturers' prediction platforms; it is currently only in one out of the three studied here.

Related note: regardless of how it is accomplished, if you are widening the dispersion of your system in the low-mid range, you are also decreasing the Direct:Reverberant ratio, which may have a perceptible impact on the perceived impact or "tightness" of the system. In the strictest sense, this is not a penalty invoked by beamsteering itself, but rather the natural results of deploying a wider directivity (lower Q) system in a reverberant environment. Regardless of the cause, it could contribute to a mix engineer's perception of the results.

Side Lobe Behavior

Recall in Figure 2 how the rear lobe moved as well as the main lobe. This is an often-overlooked side effect of electronic steering, and care should be taken to educate yourself about what is coming off the sides and back of the array, and how it will respond to the steering you're about to implement, before dumping unanticipated energy into reflective side walls or onto the stage. Although this behavior is an inherent reality of electronic beamsteering, not all loudspeakers behave equally off-axis, and not all prediction software platforms allow you to easily evaluate these effects, so in my estimation it would be fair to say that some products and ecosystems may pose a greater concern here than others.

Conclusion

In closing, this investigation helped me gain a deeper insight into what these toolsets do, and how they do it, which allows me to make a more informed decision about when or when not to deploy them in the field. As with any investigation, many questions remain, and readers are urged to explore further from this jumping-off point.

Four

Crossroads Array

An unexpected side effect of deploying novel subwoofer arrays in the field is having to come up with something to call it if it works well enough that you want to use it again.

What I have christened the Crossroads array owes its name to a festival at which I recently deployed it. The festival offered a unique combination of design parameters:

- Held in an indoor arena (flat floor with raked seating area on the sides). The primary directive was to cover the dancefloor lying in the center of the space, but anyone who wanders up into the raked side seating areas for a break should still be able to enjoy the performances.

- Twin stages, facing each other, separated by about 180 feet, with the dancefloor located halfway between the two stages.

- Mix position located halfway between the two stages, off center to both (approximately on-axis with one side of the PA). Thus, the system response at this location should be representative of most of the listening area. A slight (~3 dB) LF increase on the dancefloor may be desirable.

- A large quantity of acoustic instruments on stage, so rear rejection is paramount.

Our subwoofers cannot be flown for this event, so everything will stay on the ground. We have eight double-18 elements per stage at our disposal. They come off the truck in carts of 2. Applying some basic design theory says our line should be relatively short, since longer lines create narrower horizontal dispersion, which is not what we want here. Instead, we want a wide coverage pattern that extends well beyond the sides of the stage and spills up into the side raked seating. I decided to start by placing the sub carts evenly spaced across the front of the 32' stage width.

Since we are not pressed for real estate in front of the stage, the outer carts are physically toed out a bit to widen the coverage, and can be electronically delayed to further widen the coverage pattern to taste.

Since keeping the sub energy off the stage is important, we want each cart of two subwoofers to be deployed as a gradient (cardioid) array.

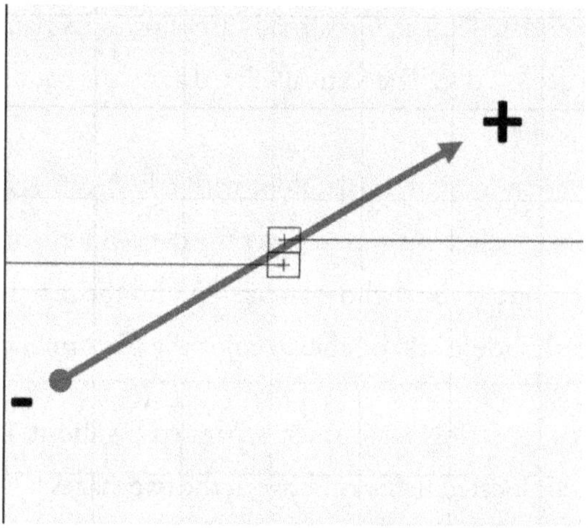

Figure 4-1

In a two-element gradient array, the vector between the acoustic centers of the two sources causes a vertical steering effect. With the bottom subwoofer rear-facing and the top subwoofer forward facing, the resulting energy will be steered upwards in the vertical plane, while the cancellation is steered downwards. This is borne out in the prediction (63 Hz).

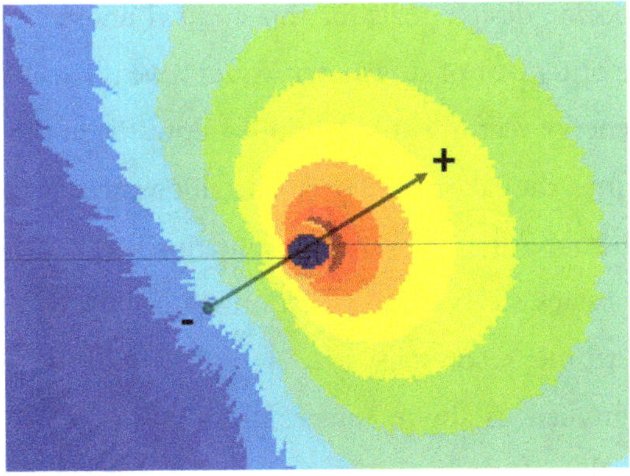

Figure 4-2

Reversing the orientation of the cart, such that the top box is rear-facing and the bottom box is front-facing would create the opposite effect (downward steering of the front lobe, and upward steering of the rear cancellation).

Let's take a look at how these two orientations perform when half-space loaded: sitting on a floor, instead of floating in theoretical free space.

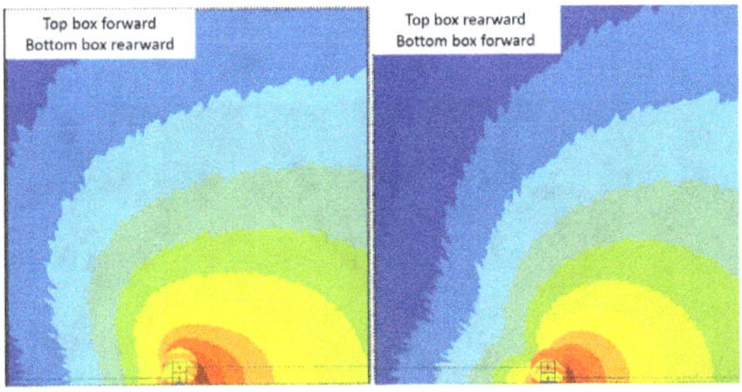

Figure 4-3

The tradeoff here is that the Front-Back (FB) orientation gives better forward-going efficiency and will better "throw" sub energy forward into the listening area, while the Back-Front (BF) orientation does about 6 dB better at placing a cancellation on an elevated stage behind the array.

We can "custom tailor" the behavior in the vertical plane by changing the orientation of each sub stack, and if we don't like it once it's deployed, it's easy to bail out — simply disconnect the cables, rotate the entire stack 180 degrees (they're on wheels!) and connect the cables again.

Given the specific design goals for this event, I chose a hybrid of the two orientations: the outer pair of subwoofer would have the rear facing box on the bottom, so the energy vector would be tipped upwards in the vertical plane, to better cover up into the raked seating incline along the sides of the room. The inner pair of subwoofers has the reverse orientation – the rear-facing box being on the top of the stack steers the cancellation vector directly towards the artists (and acoustic instruments) on stage.

The final ingredient is about 3 ms of delay on the outer carts, which has the simultaneous benefits of lessening power alley and tightening the impulse response of the array at the off-center mix position. The resulting coverage was exactly what I hoped for, with consistent coverage throughout the listening area, a very slight (about 3 dB) buildup on the dance floor, smooth coverage up into the inclined seats on the sides of the room, and an impressive amount of rear rejection on stage. In fact, the subwoofer energy directly behind the array was about comparable to the energy coming from the other stage, 180 feet away. Below is a 3D prediction and a photo of the deployment.

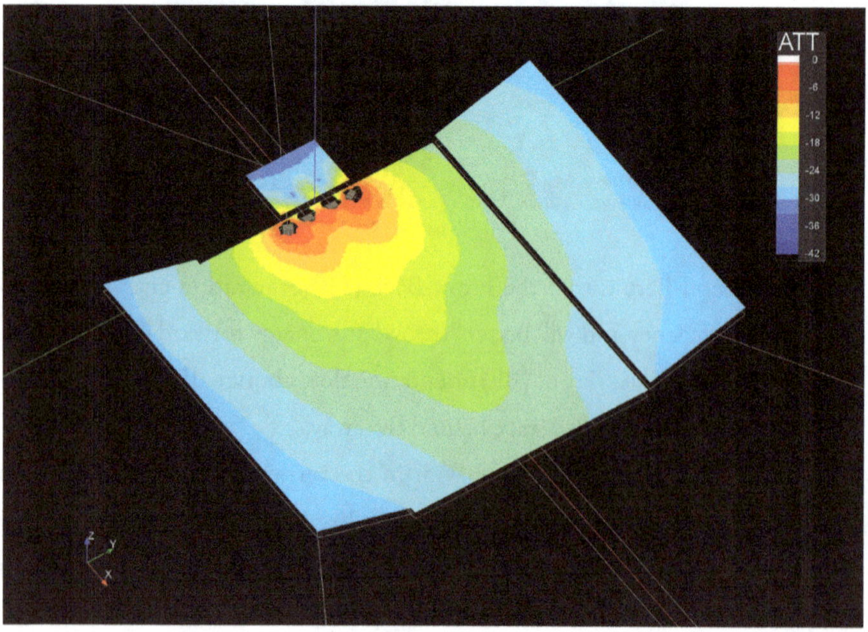

Figure 4-4

Figure 4-5

Five

Shade and a Haircut

Step by step of a basic small line array tuning

Today's job is at a local 200-cap performance venue. Our task is to remove the existing PA (4 cabinets of a last-generation small-format array per side) and put up a newer PA owned by the same vendor (8 cabinets of RCF HDL-6A per side). It's a newer, more powerful, and lighter loudspeaker and a longer line, so it's sure to be an improvement in many ways. The venue is a simple rectangular listening area, about 70 feet from the PA lifts to the rear wall. The existing subwoofers and front fills will remain in place, so we'll focus here on just the main arrays.

We can get a top trim of 15 feet, and we have eight cabinets. We can easily set splay angles to achieve equally-spaced impact points from back to front. One of the issues with the previous deployment is lack of overshoot — mix position at the rear of the room is not fully inside the HF coverage of the mains, so a balanced mix there translates too brightly for the audience. Here, we've got a bit of overshoot so that mix position is well inside the HF coverage of the mains.

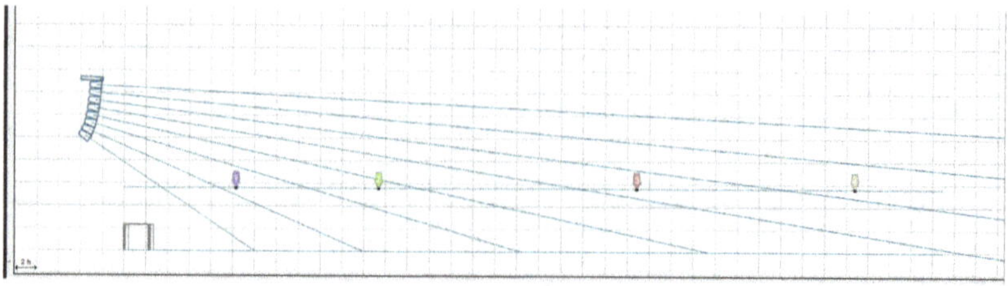

Figure 5-1

Splays are 2°-2°-2°-3°-4°-7°-10°.

We are a bit limited on DSP here, and have 3 zones available for each array, so I made the decision to zone as Top 3 (A), Middle 3 (B), Bottom 2 (C). In general it's better to have more granularity towards the bottom of the array where there's less overlap between elements. 2-box zones would be more ideal here but 3-3-2 is probably the best compromise available given the 3 outputs.

If you have read my previous book, *Between the Lines*, you know where the mic positions are going: A, B and C.

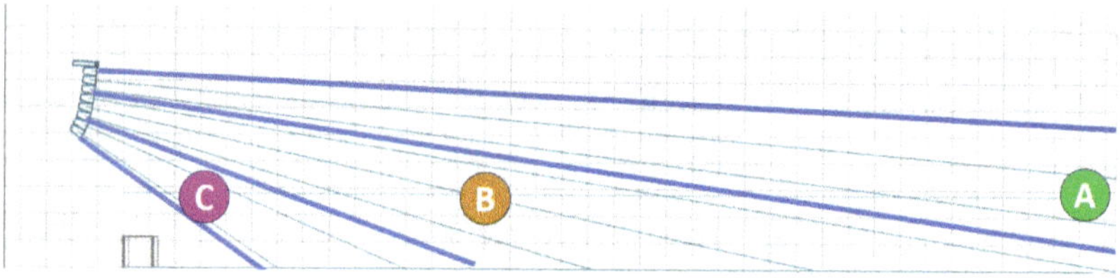

Figure 5-2

We're working with a single mic today so let's start by measuring the system's performance at each location: A (66ms), B (30ms) and C (12ms). Note that I sometimes refer to distances in milliseconds as a shorthand.

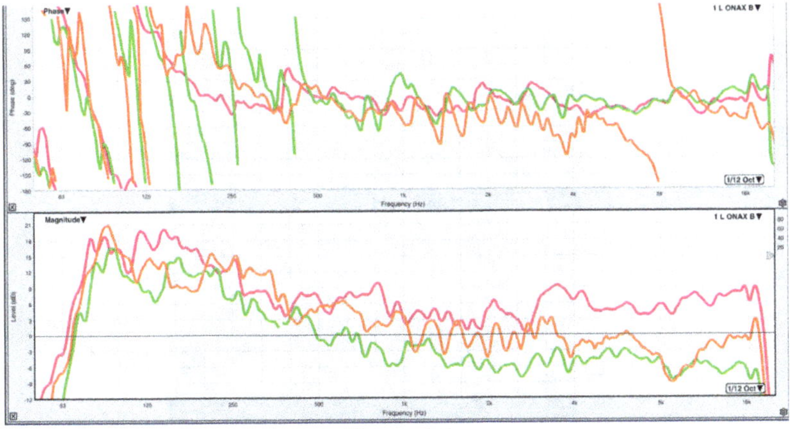

Figure 5-3

What we have here is a textbook example of array behavior before shading — we get louder and brighter as we get closer (progression from green to pink). Normalizing the traces in the low-mid range allows us to see the HF work that we have to do if our goal is tonal consistency — our bottom zone is about 4 dB bright above 2 kHz.

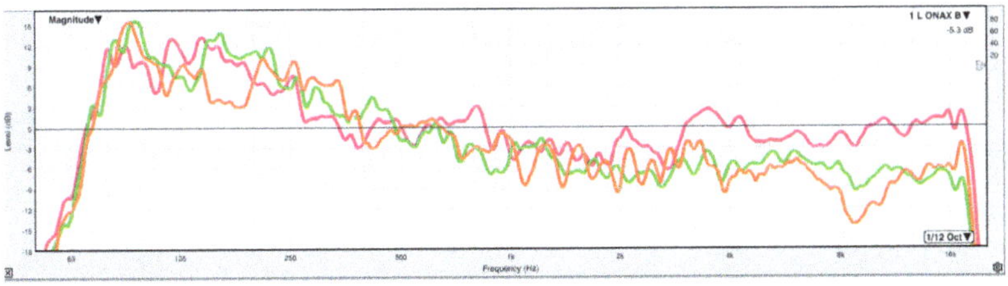

Figure 5-4

We can shave down our HF variance with some gentle HF shelving in the processing.

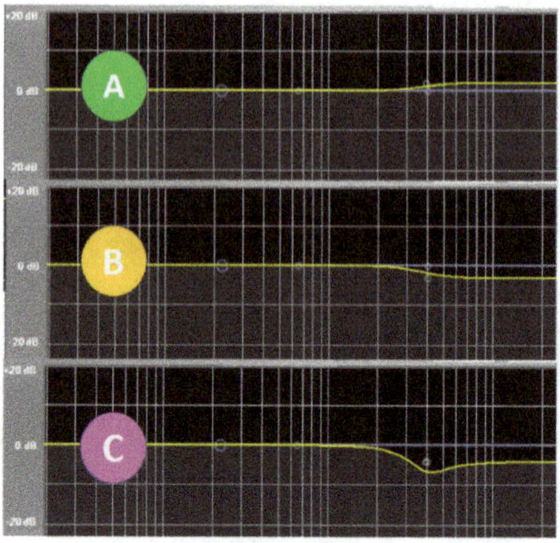

Figure 5-5

It's, as the kids say, "super effective." It's easy to get caught in the details here, but such is the nature of being in a reflective space with a reflective floor. Focus on the broad strokes and we see that the HF variance is now reduced to be more akin to the LF variance. That is, it's still louder in the front and quieter in the back but all three traces are now making a similar contour — that is — similar tonality.

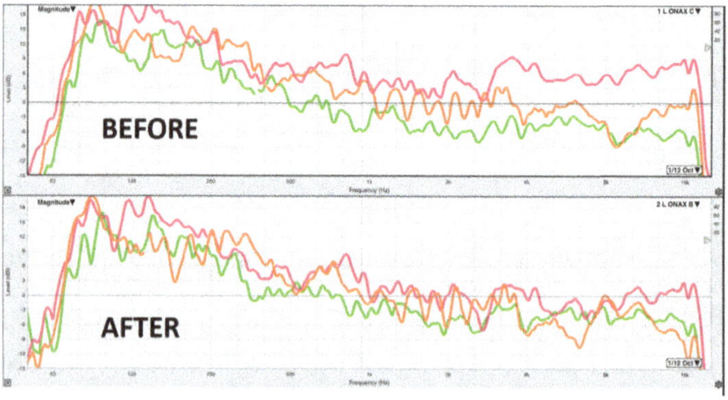

Figure 5-6

Now that we have our consistency, we will add the subs back in and do some overall EQ to hit the desired tonality for system, which in this case is less LF tilt up that starts lower in frequency. We'll do this with a combination of a low shelf cut and a PEQ cut around 200 Hz. Just a quick trim (see, now the terrible pun in the essay title makes sense.)

Notice the filters in the LF half of the spectrum are common to the entire array, while the HF filters diverge from zone to zone.

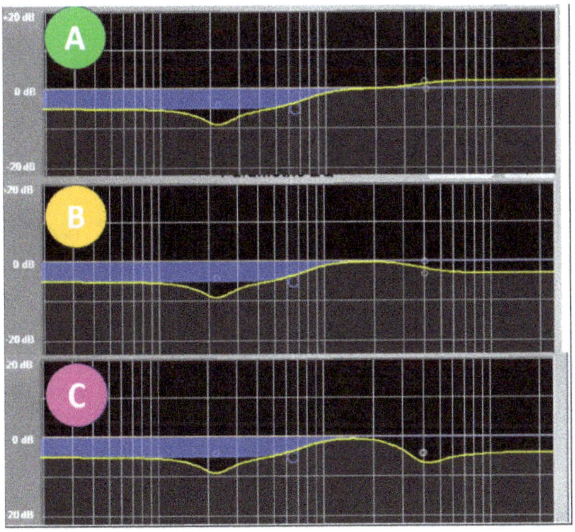

Figure 5-7

And a look at the final traces (note the different horizontal scale, now showing down to 20 Hz) with our tonality target in view (top) and the same data normalized (bottom).

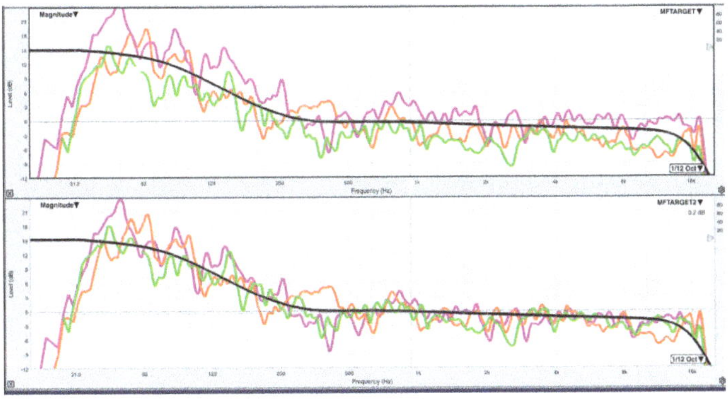

Figure 5-8

Six

Fuzz At The Bottom

A deeper dive into a concept first explored in Between the Lines

Let's examine the purpose and process of array shading and see if we can understand an often misunderstood issue.

Figure 6-1 shows the response at four mic positions on axis to the coverage of a large-format line array in an arena, before (top) and after (bottom) high frequency shading.

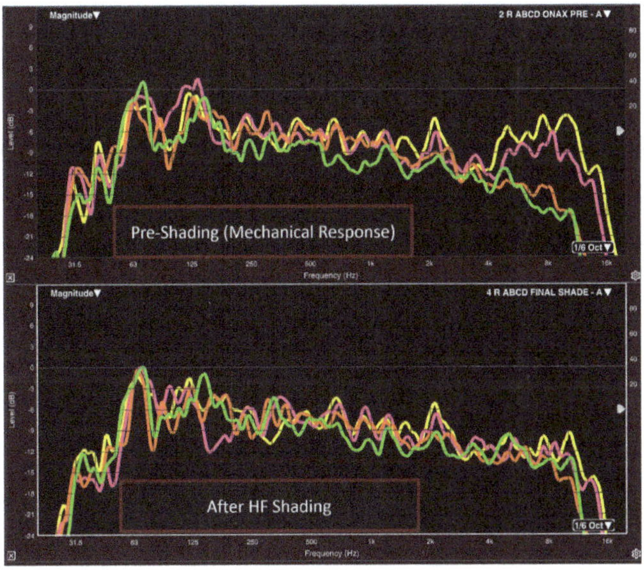

Figure 6-1

Above 4 kHz, air absorption increasingly takes effect (Fig 6-1 top). Although regular readers will recognize my familiar color coding — green, orange, pink, yellow from farthest to nearest — we can also deduce the distance that each measurement was taken from the system by looking at the amount of air absorption occurring in the high frequency range. Green is the farthest throw (168 ms), and is suffering the most loss. Yellow is much closer at the bottom of the array (72 ms) and shows much more HF as a result.

Whether applied by hand via HF EQ filters, or by some automated air loss compensation system, the goal is the same, to end up with more uniform HF from front to back (Fig 6-1 bottom). Note that the goal is not always to completely *eliminate* HF loss, and sometimes we simply want to *reduce* it, but that's a conversation for another day.

Although the above is a very successful example of shading, sometimes it may seem like it *just won't quite get there*. The extra brightness at the bottom of the array can't seem to be eliminated by EQ'ing the HF out of those boxes. The first few dB of shading are successful, and then that HF bump seems to stop responding to additional EQ cuts. What's going on?

Answering this question requires a brief brush-up on one of the key concepts in line array theory: time offset. In the lower frequencies, elements are effectively omnidirectional — that is — the energy from multiple sources is highly overlapped. (FIG 6-2, left half) Put another way, regardless of what seat you choose to sit in, you're listening to the entire array at LF. Out in front of the array, the arrivals from all the elements are sufficiently close in time with each other that the LF energy from each box combines constructively, creating a summation. (FIG 6-2, right half).

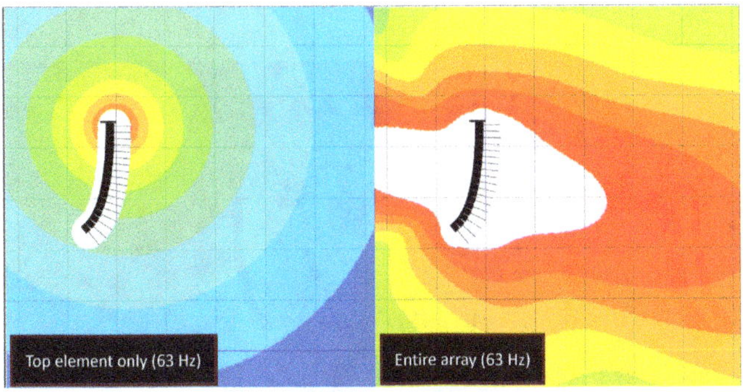

Figure 6-2

Above and below the array, the arrival times from each element are sufficiently staggered to scramble the energy arrivals such that we don't enjoy a nice, coherent summation as we do in front of the array. The longer the line gets, the more time offset we have in the arrivals and the more the pattern narrows.

If we are positioned directly above an array that's 10 feet long, for example, the energy from the bottom box arrives almost 10ms behind the energy from the top box. Longer lines mean more time offset above and below, and the LF directivity increases. Line length is therefore the major player in determining the directivity of the array at low frequencies where elements overlap their output. Want a narrower LF dispersion? Increase the line length.

This is, in a nutshell, why trying to EQ low frequencies from a single element in an array isn't fruitful, and can in fact cause unexpected results — at low frequencies, listeners are hearing the output from *all* the elements, and the whole array is riding together. We treat the entire array as one source in these frequencies.

At high frequencies, we have a completely different mechanism at work. HF directivity is created by horns and waveguides, which simply focus the HF energy in the direction we want it to go by way of pointing the loudspeaker in that direction. This works because the high frequencies have wavelengths that are short enough (measured in inches) to be effectively steered by a reasonably-sized waveguide (measured in inches). FIGURE 6-3 below shows the directivity of a single element at 4 kHz — the loudspeaker is designed to have a narrow vertical dispersion. Compare this to the behavior of the same element at 63 Hz (FIGURE 6-2) and you'll see a stark contrast.

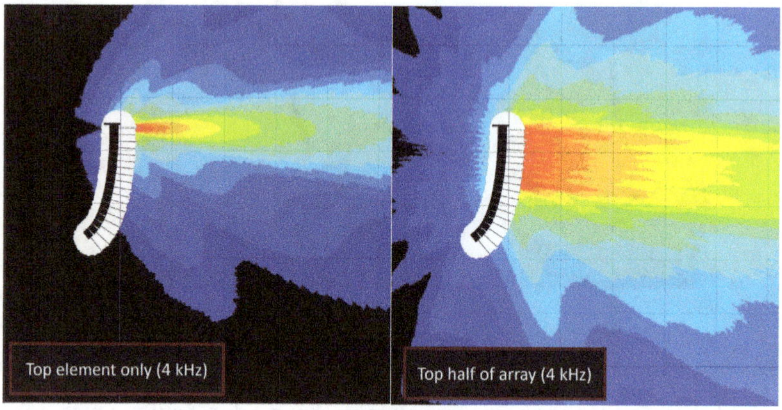

Figure 6-3

The tight vertical pattern of the waveguide creates a lot of isolation as we move off-axis from the loudspeaker, unlike what's happening at LF. To combine this with the previous concept, we could say that when listening to a line array, at LF you are listening to *all* the elements, and at HF you are listening to *only the elements pointed at you*. The waveguide is pretty narrow vertically, and once you move outside of its pattern, you're not hearing its HF output.

Since we have different elements covering different listeners at different distances, we can very effectively adjust the HF output of each element to tailor the response as needed depending how far that particular element is from the listeners it's covering. We can boost HF towards the top of the array, on the elements that will throw through the most amount of air and suffer the most HF absorption before reaching their listeners. (Of course we try to accomplish as much of this as possible via mechanical means — properly adjusted splay angles, with more overlap and therefore more summation at the top of the array. But processing will carry us over the finish line.)

The excess of HF that often occurs towards the bottom of a large array's coverage (scroll up and review Figure 1 again) isn't *just* due to less HF absorption. There's another mechanic at play here, and it's not quite as well understood. Let's see if we can sort it out.

Look again at FIGURE 6-3 above, particularly at what's going on towards the bottom of the array. It's easy to see that listeners down there on the floor are well outside of the HF coverage of the boxes at the top of the array. But that doesn't mean there's *no* energy getting down there. Horns and waveguides aren't brickwall filters, spatially speaking. They tend to do a pretty good job focusing HF energy in the right direction, but there's always a little bit of bleed off the edges. Some products fare better than others in this regard, but to some degree we're simply up against physics here.

So we *do* have some energy from the top boxes bleeding down onto the listeners at the bottom of the array, and depending on the situation, it can be more than you might think. FIGURE 6-4 below shows the output of the top half of the array at 16 kHz, using a higher resolution (1/12th octave) prediction to more clearly reveal the issue.

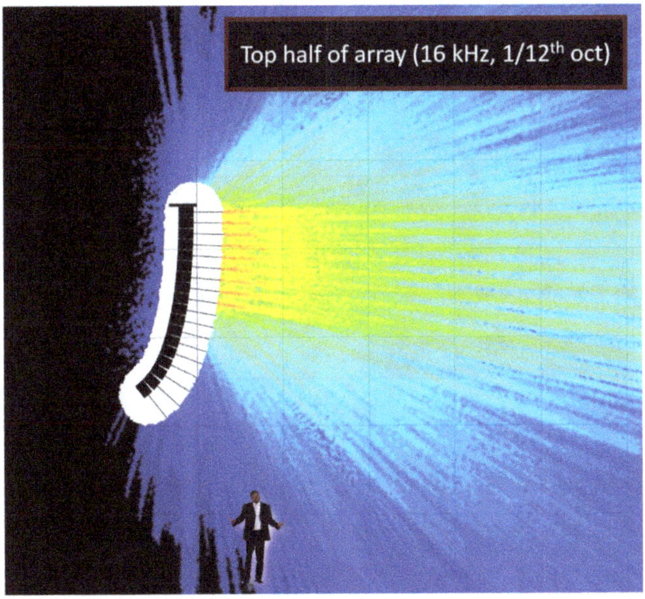

Figure 6-4

Our stock photo friend down towards the front of the audience area is definitely getting some of the "spray" — and so will our measurement mic.

Side note: This is an area where not all prediction software is created equal. The effect is not accurately characterized at lower frequency resolutions such as the 1/3 or 1 octave data used by many common prediction platforms, or if the platform isn't using Complex summation math to calculate the results. In other words, don't assume your deployment doesn't have this issue simply because you don't see it happening in your prediction. All prediction data in this post comes from Meyer Sound's MAPP 3D, which uses high resolution (1/24th octave) data, complex summation, and can produce an impulse response.

What does this "spray" look like in a measurement? We've already seen its effects in the frequency domain — extra high frequency in the magnitude response. Let's take a look at the time domain. FIGURE 6-5 below compares the impulse response of a measurement taken closer to the center of the array (top half of FIG 6-5) to the impulse response down at the bottom of the array (bottom half of FIG 6-5).

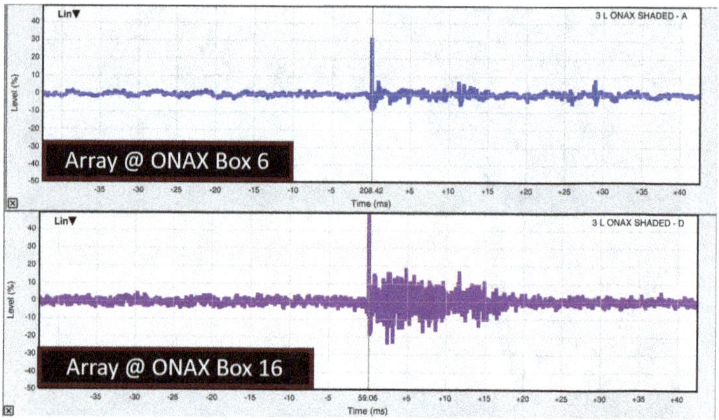

Figure 6-5

Sharp-eyed readers will recognize these graphics from the section of my book that bears the same name as this article. What's all that fuzz in the lower trace? That's the HF "spray" coming from higher up in the array. It's later in time because it's coming from higher up, further away, and therefore arrives after the arrivals from the bottom boxes. If you're trying to solve excess brightness down at the bottom of your array by shading the bottom boxes, you will only be successful to a point. The HF energy arriving later in time won't respond to your bottom zone shading, because *it's not coming from those boxes*. So your shading will reduce that initial sharp peak in the impulse response, but not the fuzz afterwards.

This type of thing can be somewhat difficult to study in a controlled way when working in a real-world production environment but prediction can bring some clarity here.

FIGURE 6-6 below shows two (virtual) measurements from the 18-box array taken from position "D," indicated by the gold star. The red impulse response shows the IR of the entire array, while the blue impulse response is just the bottom two boxes, muting boxes 1- 16. All the red "fuzz" is the late-arriving energy attributable to the rest of the array.

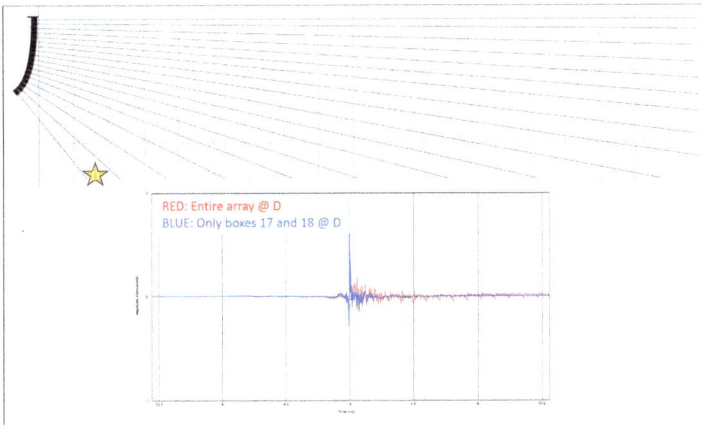

Figure 6-6

Let's get a little more granular by keeping the bottom two boxes (17 and 18) turned on, and add in boxes higher up in the array, two at a time, so we can identify their contributions to the impulse response. FIGURE 6-7 below shows the same baseline data for the entire array in red, and the effect of unmuting boxes higher up in the array two at at time in blue.

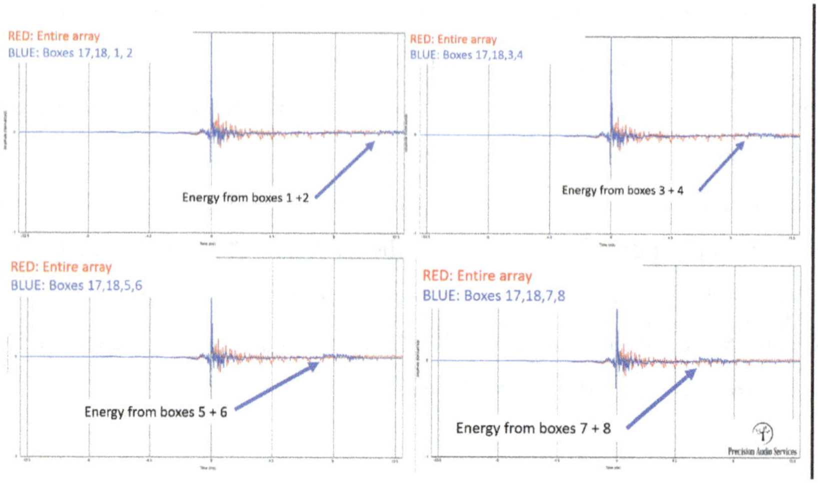

Figure 6-7

Now we have an answer to the question of why some of the excess HF at the bottom of an array doesn't always respond to shading down the bottom boxes — because that energy is coming from elsewhere.

When you turn down the HF on the bottom boxes, you make things better to a point, but now you've reduced the direct sound by a few dB, which means *more of the HF you're hearing is coming from higher up* so you've decreased your direct:reverberant ratio in an unhelpful way. This can lead to a weird listener experience down front in which it sounds like the bottom box is completely off. The ear has no strong direct first arrival to localize to.

Some people think they can get around this by leaving the bottom boxes as they are, and simply turning up the HF on the top boxes to match. You will begin to chase your tail with this approach, as turning up HF in the upper zones also turns up their contribution to the fuzz down low.

My approach — that seems to have served me well thus far — is to shade down the bottom zone to get as close to the desired target as possible, but once the trace stops responding, bring back one or two dB. This helps preserve the direct arrival, and gives the ear something to localize to, which seems to be the best compromise in terms of a listening experience in most cases. The amount of "fuzz" in the impulse response measurement from the bottom zone will give you an idea of how successful you can expect to be.

FIGURE 6-8 below shows an example of getting the bottom zone "as close as I could" — and moving on to more pressing matters. Note the Live IR fuzz for the bottom position at the top of the image.

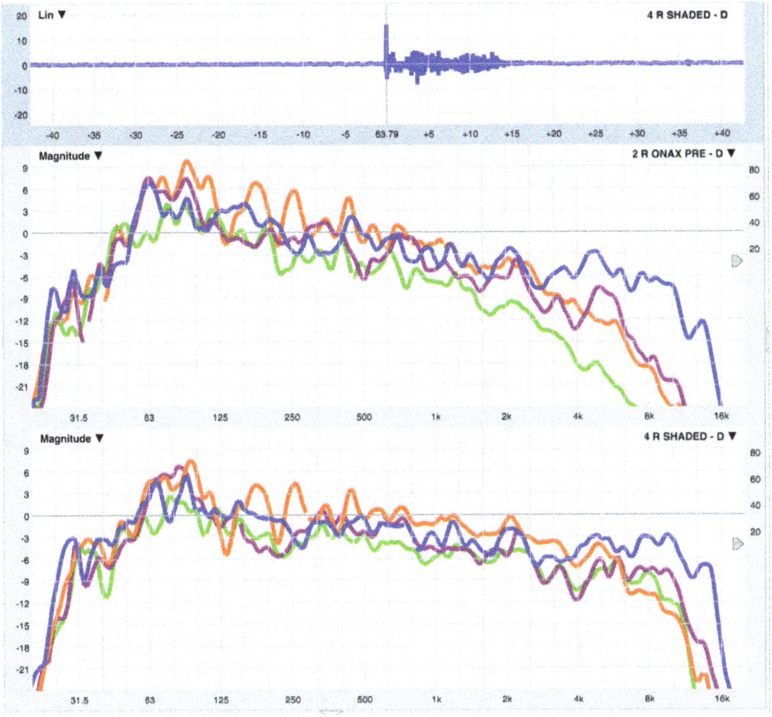

Figure 6-8

Seven

Zero Sum Game

Some insights into fill signal distribution

Let's talk about signal distribution. Some mix engineers send a Left-Right (LR) board mix to the PA, whereas some prefer to send Left-Right-Sub (LRS) and still others prefer Left-Right-Sub-Fill (LRSF). Readers hoping for me to advocate for one practice as superior in this article will be disappointed, although those interested are directed to the examination of aux-fed subwoofers in Chapter 11 of my book. Instead, I would like to take a more critical look at what I believe is one of the most commonly overlooked questions in modern sound systems engineering: how the front fills are driven, and how that decision can affect their level relative to the rest of the system.

We can begin by examining a configuration where four console matrices are provided to drive the sound system: Left, Right, Sub and Fill. This is more or less standard fare for any situation in which the FOH engineer's console is driving different PA systems from day to day.

Left and Right are simple enough — driven directly from the main stereo mix. The front fills and subs may be handled in a variety of ways, and let's focus specifically on the Fill feed for a bit (don't worry, I'll explain why as we go). There are basically 4 prevailing approaches here. The Fill feed is either:

- a mono sum of the main LR mix

- a mono sum of the main LR mix, -6 dB
- one side of the main LR mix (say, just Left)
- a completely separate mix tailored for listeners closer to the stage (say, more vocals, less drums)

The interesting thing here (and by "interesting" I mean "potentially problematic if we're not paying attention to it") is that these four methods result in *different signal levels* being delivered to the sound system.

Adding It Up

Summing the main LR mix to mono produces a signal that's approximately 6 dB higher in level than either of the contributing signals (Main L and Main R). Of course, L and R aren't perfectly correlated, so we can't expect a perfect 6 dB of summation, but in a practical sense, in the live sound reinforcement world, L and R tend to be "pretty darn correlated" because of the drawbacks of using wide stereo mixes in live venues. I tested a variety of various FOH engineers' board mixes and found that summing Left and Right reliably produced a level increase of just about 6 dB, and at least 5 dB in all cases I tested. Almost certainly, the "money channels" — lead vocal, kick, snare, etc — are panned center in the mix and therefore those elements are emerging with a full 6 dB of summation when the LR mix is mono summed.

Thus, it's reasonable to assume as a starting point that summing a LR stereo mix into a mono matrix will result in an approximately 6 dB level increase. This is the justification for case number 2, summing L and R into the front fill matrix at -6, which counteracts the effects of the summation, resulting in a signal level that matches the individual L and R signal levels.

Some FOH engineers who make use of various "stereo widening" processes in their mix wish to avoid any potentially undesirable side effects of summing those processes to mono, and so opt to drive their fills from a single side of the LR mix, say just L (case #3 above). In this case, we're operating at "unity" again, since we're not summing multiple signals together.

We can't make any sweeping statements for case #4 — a separately mixed front fill feed — since we have no idea what the FOH engineer may want to do, and so the resulting level could be different in every case.

So What?

If you're keeping score at home, this means that front fill drive levels can vary significantly from act to act, with a 6 dB difference between the two most common approaches (#1 and #2). Two reasonable questions at this point might be "So what?" and "Isn't this also true for the subwoofer feed?"

Yes, it is, but the subwoofer feed is, in a sense, "self correcting," because the mix engineer can *hear the subwoofers during the show* and can therefore make any desired level adjustments until the mix is balanced to their liking (or ask the systems engineer to do so), which is not true for the front fills in a typical situation. If the system was tuned and level balanced by sending the same (mono) generator feed to every zone in turn, which is common practice, that means the system will therefore achieve its desired level balance when the mixing console sends matched signal levels from its matrices as well. If the front fill feed for an act is derived of a LR summation with no attenuation (which for shorthand I refer to as "LR0"), the front fills are now 6 dB louder than they were intended to be when the system was tuned and pronounced balanced, and there's a good chance the engineers located at FOH can't hear it.

6 dB is considered the "upper bound" of acceptable level variance in many contexts, in which case we've lost the game right out of the gate, but there's a more sinister side effect as well: since every 3 dB level increase halves allowable exposure time under NIOSH REL, the listeners covered by the front fills accumulate four times the amount of sound exposure as listeners farther back. While folks near mix position might enjoy an hour-long performance at, say, 100 dBA, the folks down at the barricade are now enduring a crushing 106 dBA, and have reached

the same sound exposure dose after just 15 minutes.

Alternately, we could align our systems in such a way that we presume the front fills will be an LR sum, and set them to achieve the desired level when driven with the extra 6 dB. In this case, anyone driving sending either the LR sum attenuated by 6 dB (which I abbreviate as "LR-6"), or a single side of the mix, now has front fills that are 6 dB too low, which — given the proximity to subwoofers and stage bleed (acoustic drums, guitar amps, etc), is a significant enough deficiency to prompt complaints and people unable to hear the show as intended. Certainly not as concerning as the opposite case resulting in hearing damage, but both are situations we'd like to avoid. How?

Down In Front!

Of course we are responsible systems engineers and we walk the coverage during the show and adjust as necessary, but it might take us a song or two to get down there, and in many situations, front fill coverage can be extremely difficult to accurately evaluate during the show, as it requires wading through the most packed and enthusiastic section of the audience. I also like to ask the FOH engineer how they're producing their FF feeds, when the situation allows.

6 dB is a lot of error right out of the gate, so we need a method to reduce it to acceptable levels very quickly — especially in festival situations with a lot of acts, a lot of consoles, a lot of showfiles, and a lot of mix engineers sending us LRSF feeds derived in different ways.

When I align a sound system, I inject my analyzer's generator directly into the processor and route it at equal level to all zones. (Figure 7-1, below).

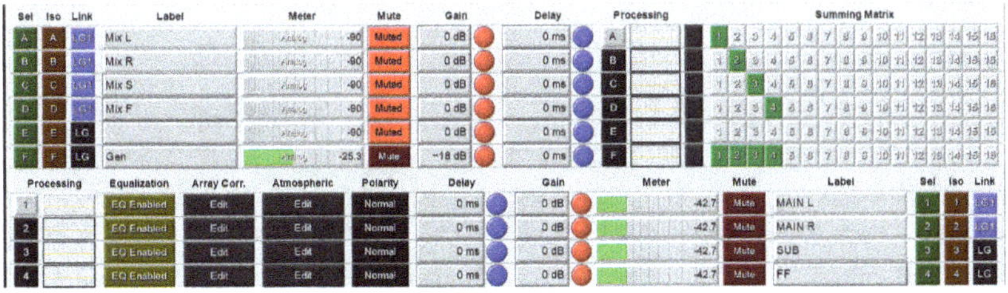

Figure 7-1

We start everyone off on a level playing field and use the output gain for each zone to balance the system as needed. (Astute readers will note the 18 dB of attenuation on the generator input, which I won't be diving into here, but is discussed in Chapter 15 of my book.) There are a number of benefits to driving the DSP directly rather than routing the generator through a console, but most relevant to this discussion is that it avoids introducing level offsets as a result of the console's Panning Law into the system alignment, as well as steers clear of all the potential routing complications and considerations inside the console.

Once we have the system tuned to our satisfaction, we have a very useful starting point: as long as all inputs to the system are at the same level, the level balance between the system zones should be preserved. Thus, when the first act of the festival kicks off and our DSP input meters look like this:

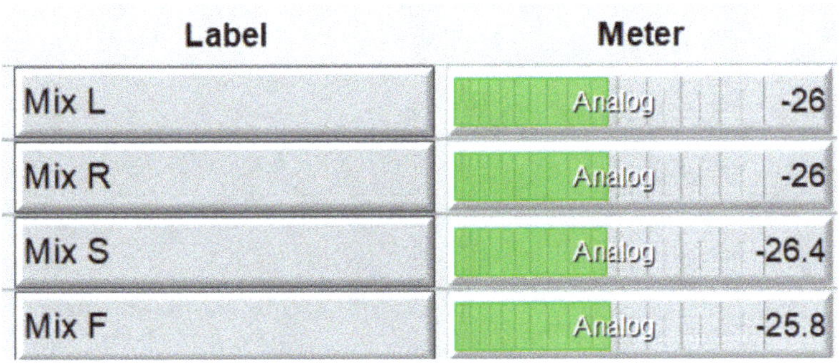

Figure 7-2

we can be reasonably certain that the front fills are in the ballpark of the correct output level since their input level is matched relative to the main LR signals. A FOH engineer sending us a Front Fill feed that is Left + Right with no attenuation (LR0) looks more like this on our DSP:

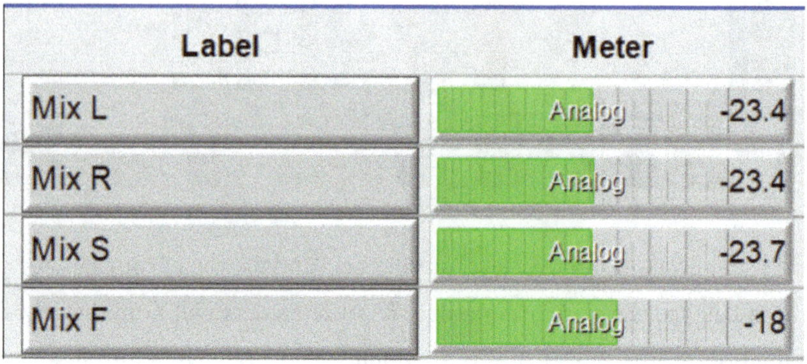

Figure 7-3

A glance at the metering tells me that I can knock the FF feed back 6 dB either on the input gain or in the routing matrix and quickly get things back closer to where we want to be, with a coverage walk to fine tune.

Note that broadband signal level on the DSP input meter isn't telling us everything, particularly in the case of the FOH engineer applying a high pass filter to their FF matrix before they sent it to our DSP — the signal level could look lower than Mix L and Mix R but still be the appropriate SPL for those listeners, so the method isn't foolproof — but it's extremely helpful when dealing with chaotic festivals for keeping things on the rails.

Eight

Roll Your Own, Part 1

Looking at bespoke subwoofer deployments

In the *Friends In Low Places* chapter of *Between the Lines,* I showed the three basic techniques by which a cardioid-type subwoofer response can be achieved using two elements. Regular readers are also familiar with the effect of increasing line length — namely, increased directivity in that plane.

I challenge readers and practitioners not to think about this work as selecting from a mental list of "vanilla" subwoofer configurations, but rather, assembling these conceptual Legos in various permutations to create bespoke subwoofer arrangements to address particular coverage requirements.

In The Crossroads Array, I showed an example of how to utilize one particular variable — in a two-box gradient stack, are we to reverse the top or bottom? — in order to direct the front radiation and/or rear cancellation in specific ways.

In this case we expand on that concept to create yet a different coverage shape for yet a different coverage requirement. Our particular application is a concert held in a ballroom-type venue that is wider than it is deep. This is always a surprisingly difficult coverage shape to create, typically approached by a broadside subwoofer line with a heavy delay arc to widen the coverage pattern.

However, in this case, logistical limitations restrict us to a Left/Right placement for the subwoofers, on either side of the 42' wide stage. Although I do not share the strong aversion to all L/R subwoofer deployments that many of my colleagues have, it is certainly far less than ideal with the particular audience geometry we have in this case, as the power alley coupling creates a narrow lobe down the center of the venue, and less energy on the sides, which is the opposite of what we want.

The first step is to create a bit of low-end rejection on stage. Since the subs will be placed on either side of the stage, a vanilla cardioid that cancels directly behind the array is not the ticket here – rather, we want to steer the cancellation off to the side and land it on the middle of the stage where the artists are.

That means the physical cabinet reversal must happen in the left-right plane, not the front-back plane. To accomplish this, I stood all four subwoofer enclosures up on their sides with the most onstage unit facing rearwards. (If you do this with 2100-LFC as I did in this case, be mindful to remove the rigging pins and "tuck" their lanyards out of the way as you stand the subwoofer up, otherwise you will crush them beneath the enclosure.) We then connecting the rigging links between the cabinets on the top side to lock the deployment together.

Figure 8-1

To predict this arrangement, rather than placing four individual loudspeakers in the software, you can save yourself some trouble by inserting it as a single flown array with 4 cabinets, reversing the bottom cabinet, and then spinning the entire thing on its side.

The conceptual vector analysis tells us that we should expect a cancellation steered onto the stage with this arrangement, and that is in fact exactly what we get.

Figure 8-2

From an alignment standpoint, I started with the delay time indicated by the manufacturer's rear-facing preset for that element, and then massaged it by hand from there. With the microphone placed on stage, I measured the three forward-facing elements, stored that trace, then measured just the rear facing element. Since the forward-facing energy is arriving from three units all further away than the rear-facing energy (and further away by different amounts), the delay needed on the rear-facing unit was slightly more than the delay value in the preset starting point.

Note: It bears mentioning that these configurations work by manipulating the relationships between the front and rear radiation of multiple cabinets, and should not be expected to work with "single box cardioid" elements which use internal rear-firing drivers to create cardioid coverage from a single cabinet — simply because the rear radiation is required for these types of arrays to work as intended, and single-box cardioid elements have significantly reduced rear radiation.

We also have an unequal fight here – we are expecting a single rear-facing element to go toe to toe against the rear radiation of the other three – it's not a fair fight. We will get a more effective cancellation – and a bit more gas out front – by increasing the level on the rear-facing cabinet a few dB to bring it closer in line with the level contribution on stage from its front-facing peers. In this case, I observed that a 2 dB level increase was necessary to level-match the forward facers.

But we must be careful here – it is often cautioned that level offset between elements in subwoofer array can result in unpredictable behavior when the rig is pushed hard, because some elements will limit before others. If you are expecting to wring every single dB of SPL from your PA, this is a venture best avoided. However, the overall output of the subwoofers was far in excess of what I needed to bring it into the desired balance with the full-range mains, and my final tuning had them turned down by 7 dB, meaning the rear-facer was operating at a gain of -5 dB. We had headroom to spare, so I am not at all concerned about the small level offset within the array, and in fact during the show, the array never even came close to full output.

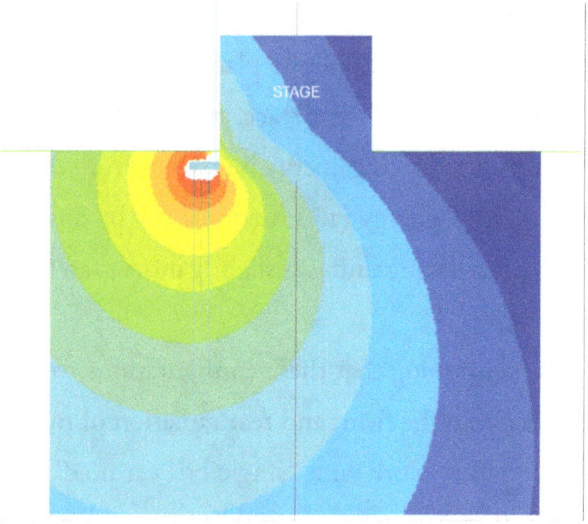

Figure 8-3

So now we see – and prediction confirms – that this half of the array is doing exactly what we were hoping for. Can't we just fire up both sides and move on to the next task? No. Due to significant overlap between Left and Right sides, the combined result is undesirable, creating a savage power alley pile up in the middle of the space and thinning out energy on the sides. This is the opposite of what we're shooting for.

Figure 8-4

To assuage this, I implemented a rather crude version of an age-old technique first introduced to me some years ago by FOH engineer Jim Yakabuski (although I have a vague memory of Dave Rat showing me a version of the technique using 5-band analog PEQ's in his FOH rack sometime back around 2011), using complementary EQ filters to alternatively emphasize and de-emphasize frequencies between Left and Right sub arrays. Although there are a number of approaches to subwoofer decorrelation, I have never been a fan of the time-domain implementations as they cause significant degradation to the array's impulse response. Jim's approach used an alternating series of high-Q boosts and cuts on his Lake processor. I implemented a lower-Q version of the same idea on the system's GALAXY, which offers a minimum PEQ bandwidth of 0.1 oct. This is just as well, as the extremely narrow PEQ boosts can disproportionately increase the resonance of certain frequencies, causing an unpleasant ringing if program material lands right on one. I also used a lower-gain version, limiting the boosts / cuts to +/- 6 dB.

If we want, say eight PEQ filters spaced evenly (on a logarithmic frequency axis) throughout the sub range, from 25 Hz to 90 Hz, we can calculate the frequency spacing coefficient by finding the 7th root of 3.6. (90 Hz / 25 Hz is a frequency ratio of 3.6, and we want to start at 25 Hz and land at 90 Hz divided into 8 filters, or 7 hops.) The 7th root of 3.6 is very nearly 1.2. Starting at 25 Hz and multiplying repeatedly by 1.2 gives us filter centers at 25 Hz, 30 Hz, 36 Hz, 43 Hz, 52 Hz, 63 Hz, 75 Hz, and 89 Hz, give or take some rounding here and there. We use same F and Q for both sides, but complementary gains: where one side boosts, the other side cuts.

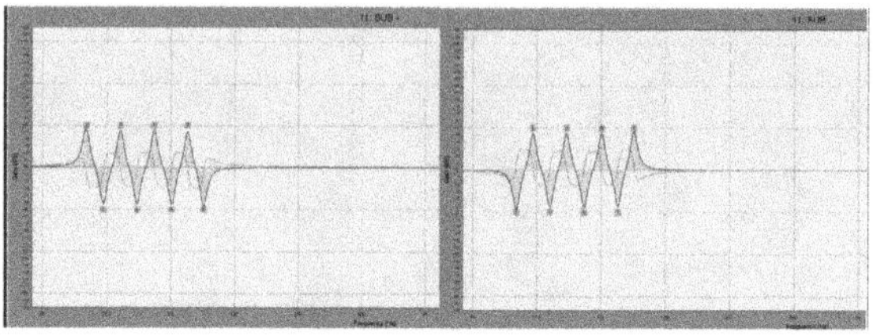

Figure 8-5

Giving one side the level advantage at a given frequency means the arrivals at the center of the venue are no longer "equal level" and so the combined summation is greatly reduced.

This is, like everything else, a trade off. Or, more accurately, several. How much gain to use is a tradeoff. More filter gain means more decorrelation but increased audibility of the filters. You have a similar call to make when it comes to filter spacing and Q (resonance). To quote the Arbiter, we trade one villain for another. And of course we need a bit more headroom in the subwoofers. The nice thing about this method is that you have it "on a lever" — it's not just on or off. You can dial down the filter gains if it is deemed too audibly objectionable, and get back a little more power alley as a trade off. (The converse is also true: you can "dial up" or "dial down" the power alley to accommodate preference.) It's also possible to disable the filters entirely if it's not worth the tradeoff in that particular space. But as we see below, it reduces the combing to a single lobe just off center, and the depth of that null has been reduced by 9 dB.

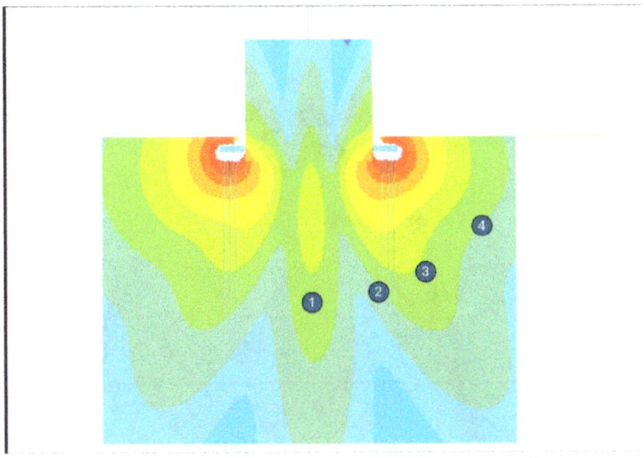

Figure 8-6, with approximate mic positions for Figure 8-7

Most importantly, there's no level buildup on center at mix position compared to the listeners seated further off to the sides — we have +/- 3 or 4 dB along any given arcuate distance from the stage — and thus our lowest octaves now make a shape that is a much better match for the HF coverage shape made by our mains and outfills. If additional width was needed, the outermost element in each array could be delayed a touch, but after listening to the deployment, I decided I liked it as-is.

This is also a prime example of "it looks worse than it sounds." Remember a prediction shows a single frequency range, and that's not how we hear. Removing the level build up on center and drastic drop offs on both sides means that listening while walking the above-described arc results in a lot more *perceptually* even coverage, as the large overall level variations from center to outside edge are significantly reduced. After all, any time we put subwoofers in a room with walls, we introduce narrow-band rippling much like these filters (intentionally) introduce, and for the most part they tend not to be overly objectionable, but of course, that's up to you.

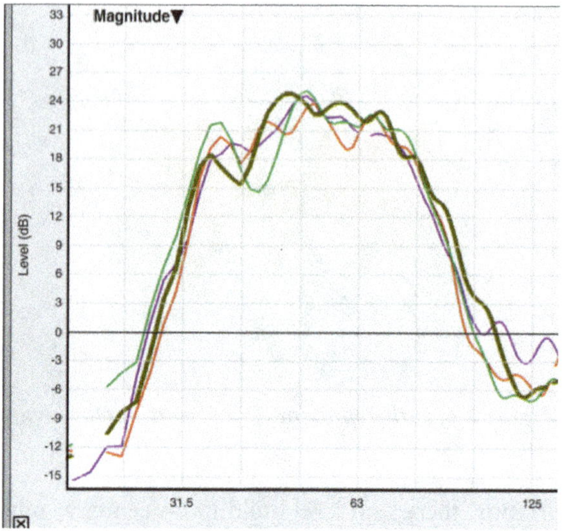

Figure 8-7

Above we see magnitude response measurements of the array with both sides on, taken along the approximate positions shown in FIGURE 8-6. No normalization, 1/6 oct smoothing. The filter ripple starts to become more prominent as the mic moves away from center. The overall level is extremely consistent between all positions. Like other techniques, it's simply one more tool in the tool kit, and it's easy to bail out of — if you implement it, and don't like the way it sounds, you can turn it off.

Also, one additional technical note — not all DSP are created equal when dealing with very narrow filters in the LF range, so it is advisable to measure your processor directly and make sure the filtering down there is doing what it says before implementing the technique in the field.

The intended takeaway here: combining a variety of established techniques and principles to create an array that meets specific design goals.

Nine

Roll Your Own, Part 2

Looking deeper into bespoke subwoofer deployments

This time, we are tasked with the design of a system for one stage of a multi-stage, multi-day electronic music festival. The system must accommodate a wide range of musical styles including a variety of DJ's, full bands, world music performances, and acts with acoustic instruments such as sitars and violins.

Our audience area is roughly square, approximately 135 feet on a side, with mix position behind, and up a slight incline. Rear rejection will be important, because we need to meet EDM audience expectations for subwoofer SPL in the audience area but not sabotage the acoustic instruments and live microphones on stage. Also, since the stage runs around the clock, the technical crew is working in shifts and camping behind the stage, and we'd like for them to be able to get some rest. So we need efficient rear rejection and high forward output.

We also have some additional restrictions in terms of real estate: the stage deck is only 3' high, and there's a 12' wide thrust in the center for a forward DJ position, so we're limited both in terms of subwoofer stack height and placement. The stage and set structure have already been designed and built, leaving a 6' high by 4' wide opening in the fascia for PA.

The initial set was designed to accommodate six boxes per side of a dual-10" line array. Given the SPL demands of the program material, particularly in the low and low-mid region, and the production goals of the client, a step up in mains power class is advised. After a series of design iterations, the speaker bay cutouts in the set are enlarged to accommodate 6 boxes per side of a dual-12" cabinet, giving an overall upgrade in SPL capability as well as LF extension and headroom, plus a slight increase to line length as a bonus.

This is an example of knowing which battles to fight – making design changes to the set at such a late stage is unquestionably an imposition on several departments, but we have an understanding of the audience experience the creative team is seeking to provide, and it's our job to steer them around potholes that would result in them being dissatisfied with the outcome, or at least make sure their decisions with regard to these matters are informed.

More front-to-back level variance than usual is helpful here, to create a "choose your own experience" environment for the attendees; those choosing to stand in the front few rows at an EDM festival have an expectation of deep, tactile subwoofer energy, but the overall tonality of the mix shouldn't be substantially different as we move towards the rear of the audience area (and mix position), so we will use the prediction to examine the amount of low and low-mid drop from front to back (a function of line length) and let the HF energy gently fall off to match with gentle splay angles and minimal shading. This gives the perpetual impression of "smooth" coverage despite the baked-in level variance.

We turn our attention now to the subwoofers; the combination of low deck height and center DJ thrust forces our hand into a L/R configuration: 6 elements per side, in three stacks of two, with a one-foot gap between stacks; we will need the wraparound paths free for optimal gradient behavior.

Reversing the bottom element in each stack would give a 1:1 ratio between front- and rear-facing cabinets, which is not optimal for rear rejection or for forward-going SPL. With enclosures of this size, a 2:1 ratio is closer to optimal. And as discussed in my last post, we don't want the cancellation to go straight backwards, but rather backwards and sideways, towards the acoustic instruments on stage (more supercardioid than cardioid). Therefore, we'll reverse the bottom element of our inner stack and our middle stack, and let both elements out the outer stack face forward. This gives us our 2:1 ratio and sets our vectors up for a cancellation onto the stage. (Simply inverting the innermost stack of elements would oversteer outwards and leave a significant power gap directly in the front of the audience area — a "power valley" if you will.)

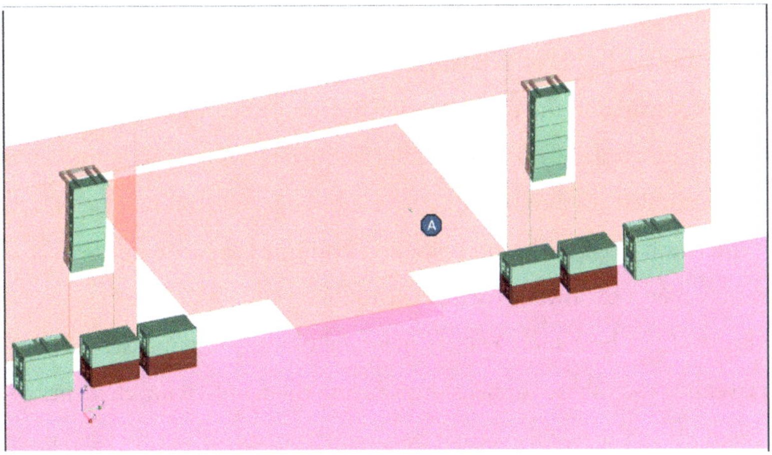

Figure 9-1

We could use the manufacturer's cardioid preset for the rear-facing elements here with some success, but we can wring a few extra dB of rejection by manually finessing the tuning, and also gain some control over the cancellation vector (back at an angle, rather than straight back). To do this, I placed the measurement microphone at about the quarter mark on the stage (approximately the mid point of *that half* of the stage, where we want the cancellation to go, labeled as A in FIGURE 1), turned on the four forward-facing elements, and captured a trace. I then turned on the two rear-facing elements and compared. The level was about 1.5 dB higher than the energy from the forward-facing elements (no doubt, at least in part because of inverse squared differences – the rear-facing elements are, on average, closer than the front-facers. I decreased the gain of the rear-facing elements to match.

Some practitioners warn against level offset between elements of cardioid arrays, and the resulting potential for "pattern implosion" if some elements hit limiting before others. However, I have found that a little level tweaking can substantially increase the rear rejection, and I did not expect the rig to hit limiting during use. (One could argue that any appreciable limiting is, by nature, an underspec condition, but that's a discussion I do not wish to dive into presently.) Consider also that numerous loudspeaker manufacturers' stock "cardioid" presets include some manner of level offset and / or EQ, and this, as far as the author is aware, hasn't killed anyone yet.

I also observed a bit of a difference in the high-end rolloff of the rear-facing elements compared to the front-facers. This is expected with physically large enclosures that increasingly occlude higher frequencies at the top of the sub range. Optimal cancellation through the entire sub range relies on matched frequency response (both magnitude and phase) so I bumped the LPF on the rear-facing elements down a bit lower to match the rolloffs.

"But wait!" I hear some readers mentally object. "The phase!" Perhaps unintuitive, but the cabinet dimensions causing the front-facing elements to have decreased HF in the rear also causes phase shift, because it's a (physical) filter, and therefore causes phase shift as well. There's no getting around it. The electronic filtering on the rear-facing elements that creates the match in magnitude response also creates the match in phase response. (Note also that the rigid set fascia creates an acoustic boundary whose effect cannot be predicted properly with typical prediction software, so measurement is required to optimize the cancellation.)

We are now well prepared to increase delay on the rear-facing elements until *both* the magnitude and phase responses match the front-facers, and then invert polarity on the rear-facers, creating the desired cancellation.

Figure 9-2

The "front and center" audience members right in front of the DJ thrust are expecting the most sub-heavy experience, and although the stage is too low for front fills, the mains have wide enough dispersion that we're not in a HF gap. We are, however, just off the edge of a now-directional subwoofer array, and the impact has lightened a bit. 1.4 milliseconds of delay on the inner-most sub stack bends a bit more energy into the center gap (think of it as a reverse-delay arc, bending coverage inward, not outward) without substantially disrupting the rear-going cancellation.

After a few hours of EDM, I felt confident that the rig would not be pushed to its limits, and took advantage of the extra headroom to apply the filter-based decorrelation described in the previous chapter, which sanded the edges off the power alley and the nulls on either side, increasing horizontal uniformity.

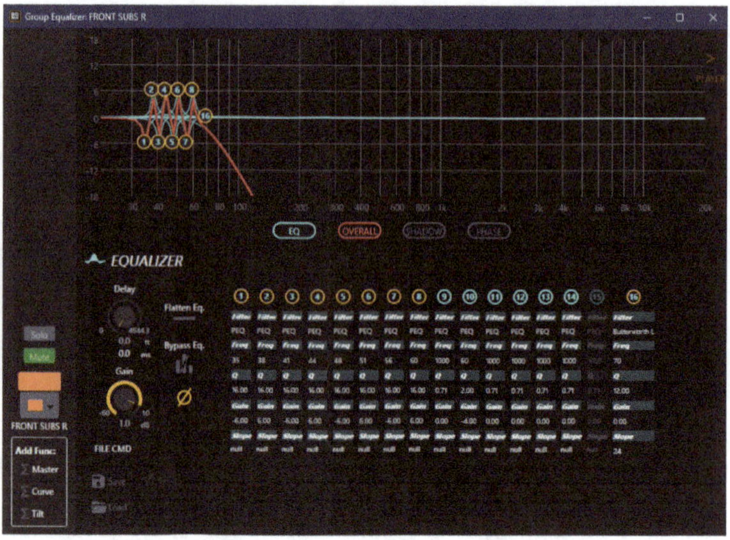
Group Equalizer: FRONT SUBS R
EQ
OVERALL
EQUALIZER
Delay
Flatten Eq.
Bypass Eq.
Gain
Solo
Mute
FRONT SUBS R
Add Func:
Master
Curve
Tilt
FILE CMD

Figure 9-3

Ten

You Shall Not (All) Pass?

Last summer, I designed a system for a festival main stage for which horizontal uniformity was a design goal. Usually it's the front to back variance that gets the lion's share of attention when it comes to uniformity, especially in many festival situations, in which systems must overcome very high range ratios (closest listeners are very close, farthest listeners are very far), often with the restriction of the relatively low trim height offered by a mobile stage.

Horizontal uniformity is often dictated almost completely by the polar behavior of the loudspeakers in use for the application. We start on-axis to the array, and as we move off-axis, HF begins to roll off until we reach the -6 dB point and the edge of the horn, at which point it's time to call in reinforcements. Our options if we don't find this behavior suitable are quite limited and typically require using a different product. Some array products allow the user to adjust the horizontal dispersion of the HF, such as with L-Acoustics' PANFLEX adjustable waveguide, offering a choice of 70°, 90° or 110° horizontal dispersion per box. The PK Sound Trinity Black offers remotely adjustable HF dispersion from 60° — 120° per module in 5° increments. d&b audiotechnik's SL-Series produces a cardioid dispersion.

This particular festival was deploying the Meyer Sound PANTHER as mains, whose HF horn exhibits a somewhat unique behavior: rather than a smooth, gradual HF rolloff between Onax and Offax, the PANTHER horn maintains more consistent HF over more of its horizontal polar, rather than starting its rolloff immediately, and then ends more quickly. Imagine your listening experience as you walk along the path of the red arrow in the graphics below.

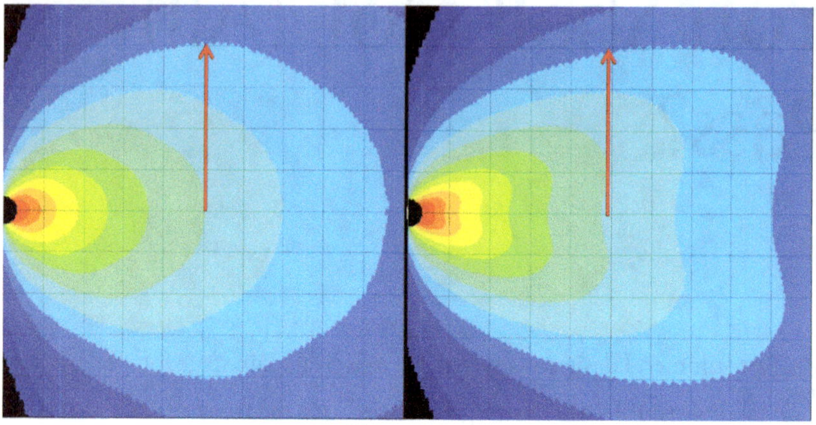

Figure 10-1

In this context, this means that we need to make sure our sub array always maintains some manner of horizontal consistency, otherwise when we move laterally, the sub array will begin to roll off faster than the HF from the mains, creating an undesirably bright listening experience for audience members towards the edges of the audience plane.

The vendor had provided 16 dual-18 subwoofers, which we deployed in carts of two as a broadside gradient array across the front of the stage. Left to its own devices, this array exhibits substantial beam narrowing in the horizontal plane, making it a poor fit for the coverage shape created by the mains HF.

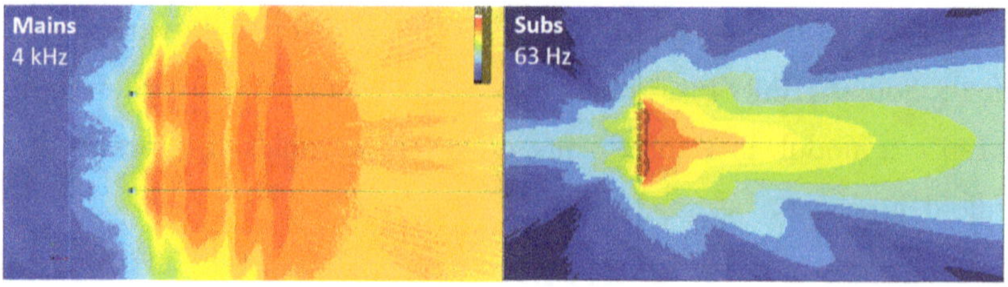

Figure 10-2

The HF extends uniformly to the edges of the listening area, while the subs have lost over 15 dB by the time they reach the outer edge. With some delay tapering, we can widen the dispersion (see Steer Me Right for more). The center four stacks ride together at 0ms, then 2ms, then 4ms on the outer stacks, a technique I've taken to calling the Pauley Array. Keeping the center 4 together helps decrease the audible smearing of the array's IR at mix position, a natural consequence of beam spreading that some FOH (including this array's namesake) can find objectionable.

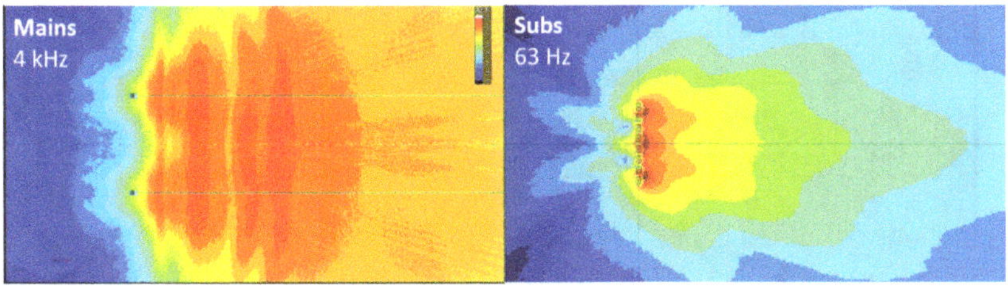

Figure 10-3

This roughly doubles the dispersion of the subwoofer array and brings it more into line with the HF coverage dispersion. This solution worked pretty well and seemed to be well received by all the mix engineers involved with the festival. However, as I am not one to content myself with "worked pretty well," I set upon improving the design for this year's festival.

These older-model double-18's have been reallocated to the B stage, replaced by 12 single-21" units. They are higher output, so we are not worried about an SPL loss, and it means we can pack that power into a smaller line length (6 carts instead of 8) which array theory tells us will give us a wider dispersion mechanically (before we start steering it).

As readers will recall from previous chapters, delay is of limited effectiveness as a steering tool because it causes different phase offsets at different frequencies, and thus the perfect amount of steering at 63 Hz is not enough at 32 Hz, and too much at 100 Hz. Perhaps we can quantify this.

The left of the image below shows the magnitude response of the array as measured at FOH, about 85' feet from the stage (red) vs the response at the same depth but 55' off center towards the edge of the coverage area. The chart on the right lists the level offset (in dB) between the two curves, or in other words, quantifies the variance between mix position and the edge. Overall the array is about 6 dB down on the edge compared to FOH, and tilted more towards the top end of the bandwidth.

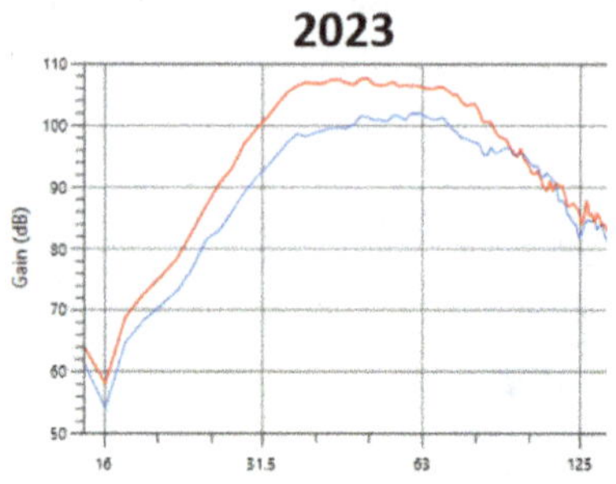

Hz	2023
31	8
40	8
50	6.3
63	5.5
80	6.3
100	0

My goals for this year were to increase horizontal consistency in two ways — both by making the subwoofer coverage shape an even better match for the horizontally consistent HF dispersion, and also by making the subwoofer coverage shape hold more consistently throughout the frequency range. Those goals could be quantified as making the FOH and EDGE curves on the above chart closer together, and more similar in shape. The new model subwoofers have better extension so we now need to be more conscious of our polar down at 31 Hz — we can't leave it as an afterthought when the rig is putting out real energy down there. The wider bandwidth means that delay is a less satisfactory steering mechanism, as the right amount of delay at 50 Hz is not enough at 31 Hz and too much at 80 Hz. We need to get frequency-dependent.

I have been asked several times since publishing Part 2 about the efficacy of using All Pass filtering (APF) for beam spreading of subwoofer arrays. The short answer is, it can be done, but — as with any beamsteering method — it is not without drawbacks. As an example, let's explore a variation of the approach to see what improvements we can bring to this year's array design — and at what costs.

After considering a number of physical deployment variations and some of the logistical limitations surrounding the stage structure, I settled on 6 carts of 2 subwoofers, with the bottom element in each stack reversed, spaced about 18" forward of the downstage edge (to give the gradient array room to work) and with one foot of space between carts (closer spacing begins to impede the wraparound paths that enable the rear cancellation and can change the necessary timings for the gradient array). Yes, this does depart from the ideal 2:1 front:back ratio for optimal rear cancellation, but optimal rear cancellation is not the number one goal here. We'll take what we get while making the desired coverage shape (i.e. a wide, even arc) out front.

I started by adjusting the delay arc until I created my desired pattern at the top of the sub range around 80 Hz. I made note of those values, and then increased the delay taper until I was happy with the dispersion at the bottom of the sub range at 31 Hz. I then input the original (80 Hz) delay values into the prediction, and subtracted these values from the 31 Hz values. The twist is to convert the excess delay to degrees of phase at 31 Hz, and use staggered all-pass to create that additional offset between the carts at 31 Hz without disturbing the top of the sub range.

For example, if 4 ms on the outer stack makes a nice shape at 80 Hz, and 9 ms makes the same nice shape at 31 Hz, we'll put 4 ms into the DSP, and the 5ms difference translates into 5 ms x (31 Hz / 1000 ms) x 360° = 56°. We would then insert the DSP's all pass filters on each sub cart, tuned to higher f in the center of the array and lower f at the ends, such that the *difference* in phase between the middle and the end at 31 Hz amounts to 56°. Yes, this is just a by-hand version of the Low-Mid Beam Control "Beam Spread" algorithm, done with a bit of trial and error and a little help from our friend Excel.

My array ended up with the inner carts at 0ms (time zero reference), the middle carts at 1ms, and the outer carts at 4ms, with 2nd order APF Q=0.6 tuned to 20 Hz on the inners, 17.3 Hz on the middles, and 16.5 Hz on the outers. Intuition might lead us to think that such small differences in center frequency would produce a negligible effect, however once we deployed the array, the FOH mixer and I did extensive listening with this arrangement vs a more traditional delay-only taper and found that the delay+APF version was significantly more consistent horizontally. I used the decorrelation EQ filters mentioned in previous posts to further smooth out some of the combing between stacks, and after another round of listening we found the result to be quite successful.

The first question to answer is: does it work? The short answer is "yes."

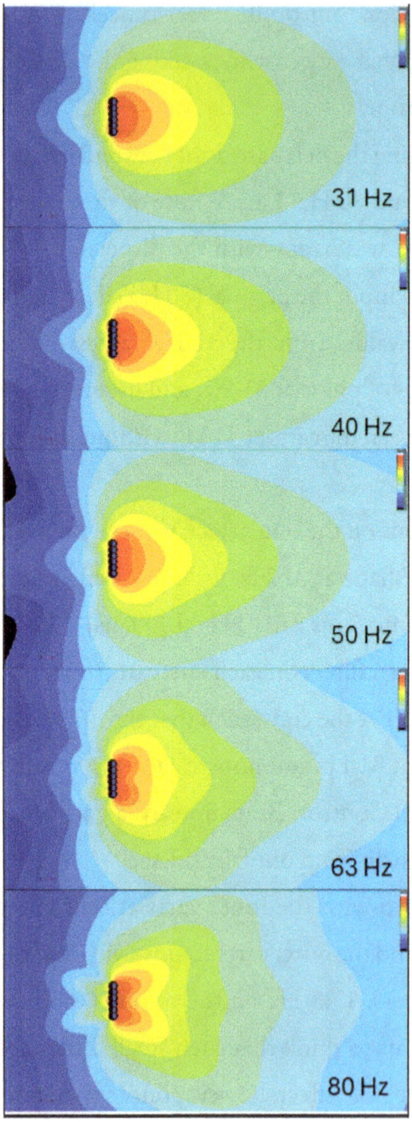

Figure 10-5

Our horizontal coverage pattern is much more consistent between middle and edge — a much better fit for the high frequency shape we're trying to match. If we return to our earlier "horizontal consistency" metric of comparing the dB loss between center and 55 feet off center at 85 foot depth for each third octave band, we can see we have a substantial improvement (red line) over last year's design (blue).

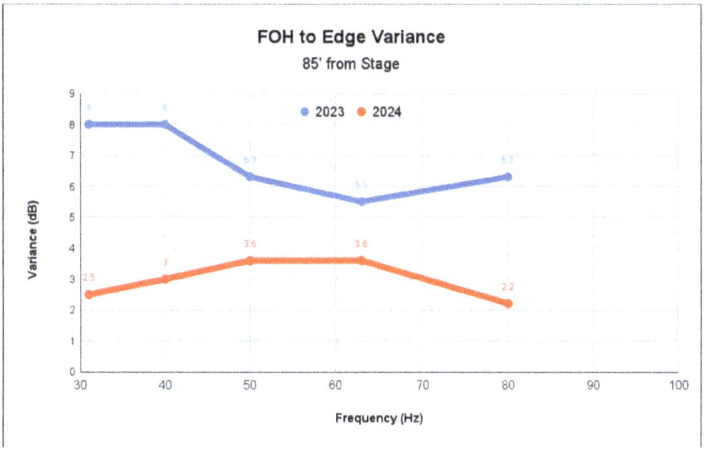

Figure 10-6

The caveats: all beam steering costs a penalty in impulse response smearing, but APF steering + delay steering costs you a double penalty. The delay pushes the stacks of subs apart from each other in time, and the APF pushes the *frequencies in the sub range* apart from each other as well, giving you in essence IR smearing in 'two dimensions.' Whether or not this is acceptable tradeoff sonically is highly situation dependent, and should be carefully evaluated before making a decision.

Note also the higher level of precision required to achieve the desired results — physical positioning, delay times, and filter frequencies need to be carefully checked in order to make our real-life array look like it did in the prediction. If production needs us to move a sub cart a few feet later on to accommodate a concrete ballast block, our array may respond in highly unpredictable ways. Also, every subwoofer in the array ends up with a unique combination of DSP settings, so patching must be carefully checked.

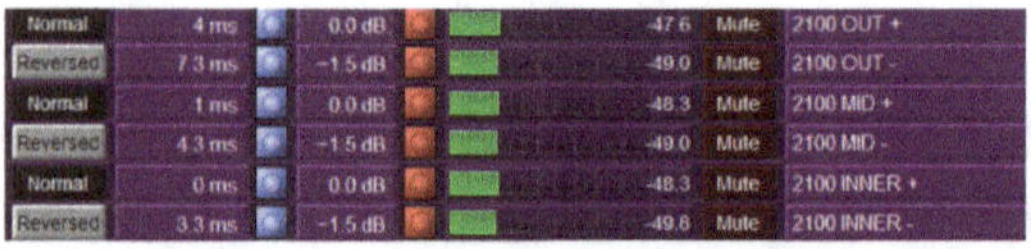

Figure 10-7

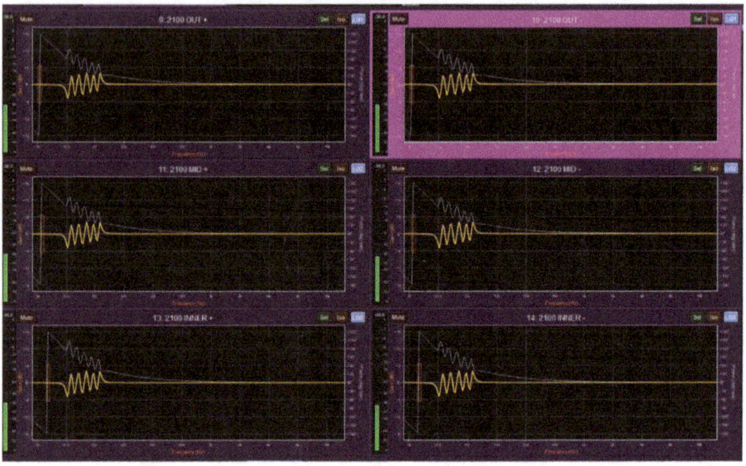

Figure 10-8 - The processing settings for one half of the array

I was quite satisfied with the end result, and it seems to be well received by the guest mix engineers (although this may be one of those "don't show them what you did to it" situations, and if any of them are reading this now after having mixed on it, I suppose the cat's out of the bag). I found the IR of the array at FOH to be well acceptable as a trade off for the wide, smooth coverage dispersion, especially given that a large portion of the audience attending the headline set was listening from the far edges of the audience area. However, given a different day, different artist, different mix engineer, different environment, etc, the tradeoff may not be the right one, and as the Bard of Avon tells us: ay, there's the rub.

Eleven

Taming Red Rocks

In the summer of 2023, I was hired to design, deploy and tune a sound system for a concert at the Red Rocks Amphitheater in Morrison, Colorado, a venue notoriously unique both for its acoustical properties and its challenging coverage requirements. Having never worked there before, I called a number of my colleagues who had mixed shows there, and asked what they considered to be the biggest challenges from an audio perspective.

I was treated to a large variety of anecdotes about various attempts throughout the years to cover the iconic 309 foot, 22° sloped seating area – custom rigging brackets, new roof trusses, flying dV-DOSC on *top* of V-DOSC. If this wide variety of system design attempts and iterations were to be catalogued in a coffee table book for sound engineers, it'd be a best seller. *Note to future self.*

Despite the variation in products and approaches used, two challenges were recurring themes: 1) getting sufficient LF and lo-mid energy up to the back of the slope, and 2) mix position is so close to the stage that it's "off the edge" of the PA coverage, located in the center gap of light HF coverage between the main hangs. Fair enough – addressing these two issues would be my twin design goals.

The key to successfully projecting LF over long distances is simple enough, at least in theory: line length. Lengthening the line will increase the directivity of our array at low frequencies, a principle well established by Harry Olson almost a century ago, and dating back even further if we look to literature applying the same principle to optical physics.

FIGURE 1 shows a "landing strip" cut from the seating area prediction, showing the effects of increasing subwoofer line length on the front to back level drop at 63 Hz. We begin with a hang of three 18" subwoofers in a cardioid (FBF) configuration. Our prediction is showing us 3 dB / color division, which translates into a 15 dB drop in LF level from front to back.

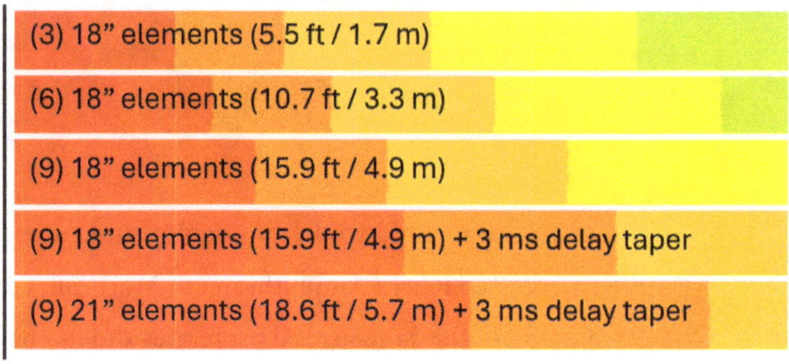

Figure 11-1

This illustrates a common limitation in large-format sound system design, and it's not exclusive to Red Rocks: some vendors and designers specify subwoofer quantity based on the SPL that must be produced at FOH mix position, and today's modern double-18" subwoofers are incredibly powerful beasts, such that three per side might be totally sufficient to create the desired SPL at mix position. But with a vertical line length of only about 5.5 feet (1.7 meters), it's virtually omnidirectional in the vertical plane, and when contrasted with our much longer line array (which is in turn responsible for projecting generally shorter wavelengths), our sub array just can't match the directivity of the full-range mains, leaving us with a system that loses sub frequencies over distance much more rapidly than the upper 6 octaves.

Modern line arrays have no trouble projecting high frequency energy hundreds of feet (atmospheric considerations aside), but if the bottom of the spectrum isn't keeping up, our last-row listeners are subjected to an unpleasantly bright, thin presentation, lacking impact.

Thus, our subwoofer line length needs to be increased to improve directivity and allow it to fight off the level drop over distance, remaining well-paired with our full-range sources. Increasing to six (11 ft / 3.3 m) is still not quite satisfactory, but a bump to nine elements (line length of just under 16 ft / 5 m) and tipping the entire array upwards by 5° bumps the green isobar out of business entirely.

We can also gain an additional 3 dB improvement with the introduction of some gentle delay taper (1.5 ms on the middle cardioid cell, which is elements 4,5 and 6, and 3 ms on the top three elements) and we have something quite respectable – a last row that is only 9 dB below mix position at 63 Hz, despite being over 250 feet further up the hill.

There's nothing novel here – the line length mechanic is doing exactly what theory tells us it will. So we will specify a higher subwoofer count, and when it is pointed out that we don't "need that many, they are plenty powerful," we will agree, and respond that we're trying to make a shape, not just an SPL, and we will probably end up turning down the gain a bit.

But our limitations here are practical in nature. A flown hang of nine subwoofers is heavy (we will need two-ton motors to fly it safely), we can only achieve the desired 5° of mechanical tip-back, and no more (this much tilt lands all the weight of the hang squarely on the downstage rigging point, preventing us from reclining the array further), and of course, someone far above our pay grade needs to approve the rental cost on the increased subwoofer quantity, the vendor must have sufficient quantity available, the truck space to bring it in and out, the time and labor to deploy it, and so forth. In that sense, the real-world limitations may mean the party is cancelled before it's even started. In this particular case, the bean counters understood what I was asking to do, and agreed with the intent, so the request was granted.

Okay, we've tackled the "LF not making it up to the rear" issue fairly simply (all the line length money can buy!), but the remaining challenge – the light HF coverage at mix position – requires a bit more nuanced approach.

How do we close up a center gap in HF coverage? We have three main tools: aim the loudspeakers inwards instead of pointing them straight ahead, move the Left and Right hangs closer together, or use loudspeakers with wider HF dispersion. Let's do all three!

Luckily, the show in question has a relatively narrow stage footprint, and no upstage video element, so we have a lot more leeway than usual with our sightline considerations, allowing us to move our main points one beam further on stage than where the PA is usually hung, diminishing our center gap by 16 feet.

We can use a bit of trig (if we paid attention in high school math class) or a right triangle solver (if we didn't) to calculate how far onstage we need to shift our downstage rigging point to add a 5° toe-in on our main hangs. Nothing in life is free – and this toe-in creates some under-covered seats on the sides in the first eight rows or so, so we will use a ground-stacked outfill array to restore coverage to the front corners of our seating area. This is, in my estimation, far better than leaving a hole at mix position. If the FOH engineer is mixing off the edge of the horn, they are hearing less HF than listeners on-axis to the PA, making the show sound bright for everyone else. Compound this with the LF loss due to a short sub line, and you have a perfect recipe for an unpleasant last-row listening experience. Restoring proper coverage to mix position should prevent this.

Finally, the vendor has four wide-dispersion elements available, so we'll place those at the proper positions in the array to cover mix position, two per side. The wrinkle is that they want to pin the arrays to the proper angles at the shop the day before, so they're loaded onto the truck ready to fly, so we need to figure out exactly where those elements need to go in advance.

I used a Google Earth satellite photo to measure the lateral distance from the downstage edge to mix position, and then superimposed a vector of the same length onto the prediction, indicating that elements 12 and 13 in the main hang should receive the wide elements. Obviously, this approach is somewhat lacking in precision, so the first thing we will do once we arrive on site is to take measurements and confirm that the model matches reality, and that our predicted array geometry is correct.

It can be a bit of a tightrope walk to go into a new venue and do things the house crew hasn't seen done before. When the head rigger, who has hung hundreds of PA's off the same roof, asks you if you're *really* sure you want to toe the PA inwards, because a lot of people choose to do the opposite...you can't help but wonder if you're about to drive straight into a pothole that everyone else saw coming and avoided. Do you really trust your design, prediction, and math skills enough to override the concerns of the technical minds who have an intimate knowledge of how to stage successful events in the space? Am I about to do something that everyone who works here knows not to do? This is probably the most challenging moment in the entire process, when we need to make a decision and be saddled with the consequences of it.

In this instance, our trust in the design process was not misplaced, and front to back measurements of one side of the PA (Main & Sub) reveal that we've met our level variance goals, showing a drop of about 5 dB from mix position to about 200 feet up the hill (FIGURE 2, top). Normalizing the traces shows that our full-range tonality is intact all the way down into the sub range (FIGURE 11-2, bottom).

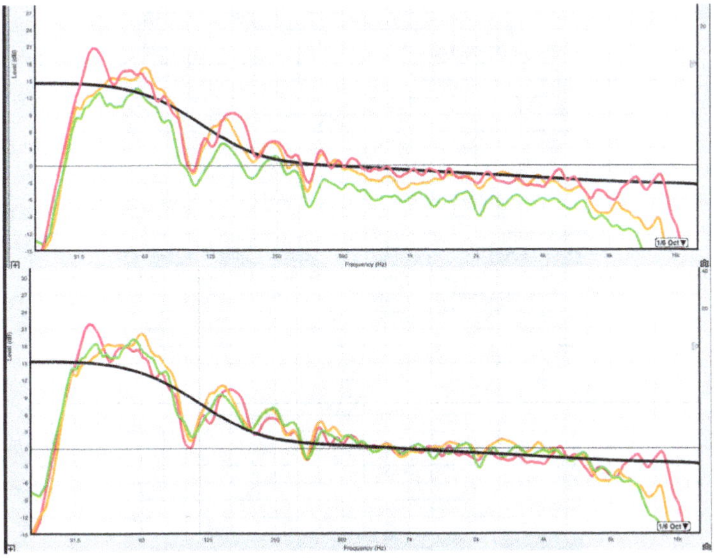

Figure 11-2

(Interestingly, I observed about 2 – 3 dB *less* LF loss in the last rows of the venue than the prediction indicated, and ended up slightly sub-heavy in the rear, leading me to reduce the electronic beamsteering a bit on the sub array. My working theory is that this is because the PA is not radiating into free space, but rather has the boundary of the rigid seating surface as well as the rock walls on both sides, which form three walls of a rudimentary "horn" to funnel the energy up the hill and slow the inverse-square loss rate.

Note that I did end up acoustically simulating this and the results were quite interesting, but that is reserved for Volume 2 (the impatient can find the article on my professional website).

Our onstage hang points, toe-in and wide elements have properly combined to ensure that mix position is in fact properly covered, and once we deploy our deck-stacked outfills, we have restored our horizontal uniformity as well. It's a bit of exercise to climb all the way to the last row, but it's worth it to sit down, catch our breath, and feel the impact of the low frequencies coming from 300 feet down the hill.

Figure 11-3 - The 2023 deployment

Figure 11-4

Anyone familiar with my design approach will know that I never pass up a chance to spin up another design iteration if I feel there's still potential for improvement, so when the same artist asked me to return to Red Rocks and handle their show this year, I gladly accepted.

We focused our improvements on four main points: first, the vendor offered a 17th box per side for our main hangs – effectively "the same geometry but with one more box at the top" to better cover the ADA seating plateau that is elevated above the last row at the rear of the venue. This proved a bit of a challenge mechanically, as this requires more uptilt on the rigging frame, and virtually all the weight of the array was already on the front point. By massaging splay angles a bit and adding a bit more curvature to the bottom of the array, the bottom boxes act almost as a counterweight for the top ones to pull the center of gravity back under the rigging frame. This gave me about 7° of over-coverage past the front row, so I had planned on simply muting the bottom box. However, the "pit" area ended up being filled with an additional row of removeable and ADA seating, a happy accident, meaning the bottom box landed right on target.

Figure 11-5 - 2024 deployment

Second, the vendor had an increased inventory of wide-dispersion elements available, allowing us to make the bottom 6 elements in each array all wide, which extended coverage further into the under-covered corners, reducing the amount of territory that needed to be covered by the outfills.

Third, we used a larger-format element for our ground-stacked outfills. They were being driven heartily the previous year, and this year I wanted to increase the splay angles at the top of the stack to create a softer HF edge and help "feather" the outfill coverage into its seam where the mains take over. A more powerful loudspeaker with a larger horn gave us the necessary output to do this with ease, and created a better tonal match with the large-format mains.

Finally, the vendor had upgraded their subwoofer inventory to a newer, single-21" model. I still had nine per side to work with, but the larger driver meant a taller cabinet height, which yielded about a 15% increase in line length, and a corresponding improvement in vertical directivity and therefore front to back consistency (see the bottom-most bar in FIGURE 1). Eagle-eyed readers will note that the LF is now starting to *decrease* in the front few rows, making front fill subwoofers important. Accordingly, our deck-stacked outfill arrays were supported by two-per-side double-18" subwoofers in gradient configuration, restoring some impact to the front rows while maintaining a quiet stage – a priority for this artist.

Figure 11-6

I will readily admit that the toed-in mains can look a bit "wrong", particularly when viewed from off-center in the curved seating area, but I am firmly in the "It ain't stupid if it works" camp on this one, and already kicking around some ideas for my next incremental revision next year if I am fortunate enough to be invited back once again.

Twelve

Saturday In The Park

A design walkthrough of a large-area coverage scenario

This time, the task is to design a sound system to cover the 19-acre Gas Works Park in Seattle WA for an open-to-the public fireworks event. The park is on the National Register of Historic Places, the last remaining coal gasification plant in the United States, converted into a public park in 1975. The plant's long-retired, massive steel tanks and pipe structures, relics of early 20th century industrial engineering, form the architectural centerpiece of the park. The sound system will radiate outward from the architectural fence line to cover the audience areas surrounding the structure.

Our event will require equipment spread over a large area, and due to the terrain, most of the gear will need to be pushed into place by hand, so we want to be efficient with our equipment choices to minimize weight and crew exertion.

The majority of the stated 50,000 person attendance capacity will occupy the lawn to the west of the plant structure, where the terrain rises to a hilltop about 250 feet away. On the east side of the plant, a paved area hosts food vendors, and beyond, more seating area on the lawn sloping down to the waterline about 270 feet out.

To the south, we have a relatively flat VIP seating area overlooking the lake, and additional VIP seating along the lakeshore to the southwest. The TV broadcast stage in between the two VIP seating zones needs to be out of coverage so as not to interfere with the broadcast. In addition, the entire north lawn will remain uncovered by our system, to accommodate a small mobile stage with band performances throughout the day. (As an aside, we were asked to accept a board feed from that stage, as well as send a feed from our system to their mixer. Care must be taken at both ends of the signal chain to ensure that neither operator can accidentally create a giant feedback loop. Fellow systems engineers are invited to consider how they might accomplish such routing.)

We will apply the design strategy outlined in my book – use the largest / most powerful system to cover the largest audience area, then use fill systems to fill in the gaps. Rinse and repeat. Here, our mains are 32 elements of a small-format line array cabinet. The West and East coverage areas and throw distances are about equal, so we will use half the boxes for the west side and other half for the east side, allocating 8 elements for each of the four main PA tower locations – NE, SE, NW, and SW. (If one side required a significantly further throw than the other, we would allocate the inventory accordingly.)

To make the most of our set up and load in time, and help ensure the vendor doesn't pack too much or too little equipment on the truck, we want to have our design as solidified as possible before we arrive on site. Google Earth Pro satellite info can give us a bird's eye view of the area, and even preliminary elevation data that we can use to translate into rudimentary audience planes. We can start by marking out on-axis "landing strip" elevation data for each of the four PA tower locations (FIGURE 12-1). The two West arrays will throw uphill, while the two East arrays are throwing slightly downhill. How high should we trim? The 25' maximum height of the lifts makes the decision for us. A pair of dual-18" subwoofers at the foot of each tower will supply the LF extension.

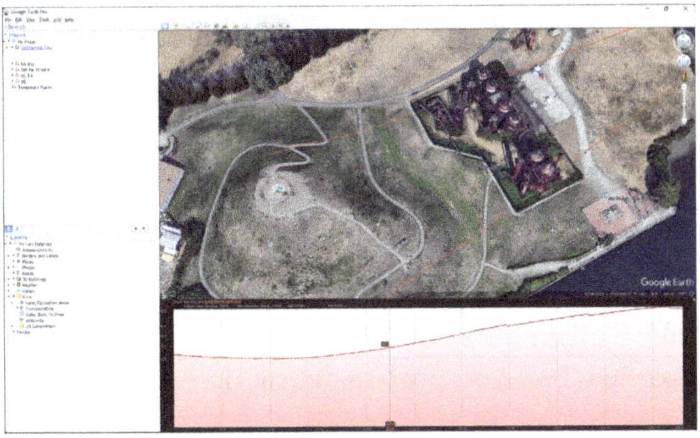

Figure 12-1

Here we must pause to examine a shortcoming of the simple on-axis "landing strip" elevation method for determining splays. Our array has a 100° horizontal dispersion, and the summit of the hill, which is about 15 vertical feet higher than the on-axis point, lies off axis horizontally. If we are only considering the ground elevation of the on-axis landing strip, we will miss covering the summit. We need to overshoot the on-axis geometry to make sure the listeners on the higher ground off axis are properly covered.

Figure 12-2

FIGURE 12-2 shows the satellite imagery from the perspective of standing at the base of the NW PA tower, and the red line indicates the on-axis landing strip elevation that we might traditionally shoot with our rangefinder and input into prediction software. The green line shows the additional vertical height that we must achieve with our coverage in order to properly cover the people at the top of the hill (who arrive a full twelve hours early to claim this "perfect viewing spot" for the fireworks show). When only considering the on-axis landing strip splays, the coverage seems to be excessively overshot, but considering the entire 100° horizontal dispersion of the cabinet, we see it's there for a good reason.

A similar issue can arise with traditional side hang / outfill geometry in arenas and stadiums, when the array is projecting onto the audience plane at a non-orthogonal angle. FIGURE 12-3 is a generic stadium side seating plane. Normally we would overshoot by a box or two to prevent the HF rolloff occurring in the top rows, but I've aimed the top element precisely into the top seating row for clarity here. The prediction reveals the severe under coverage situation at the edge of the horn nearest the array. To rectify this, we need more overshoot than would otherwise seem necessary. Using 3D prediction software is the best way to protect against this type of under-coverage error, as it is not readily apparent in 2D landing strip predictions whenever the source is not impacting the listening plane head-on (what physics-inclined individuals refer to as a "normal" angle).

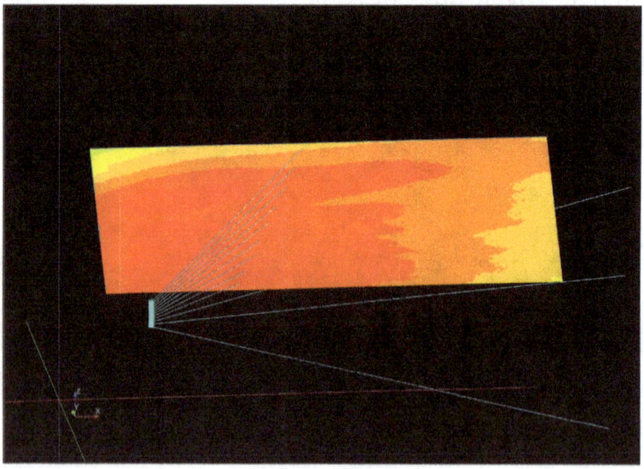

Figure 12-3

Since we have almost 200 linear feet between the NW and SW towers, we can expect extensive center gap along the fence line between the arrays, although they converge and overlap at distance, and prediction confirms this (purple shaded area in FIGURE 4). Since we don't need to go very far, we can fill this gap with two point source loudspeakers on sticks located at the two dogbone bends in the fence line (blue icons in FIGURE 12-4). Since a secondary fence around the architecture in the center provides a buffer zone area open only to production crew, we don't need to worry about people in the immediate nearfield getting blasted, thus these auxiliary fills are only tasked with covering from approximately 12' to about 50', at which point their coverage feathers into the energy from the main arrays.

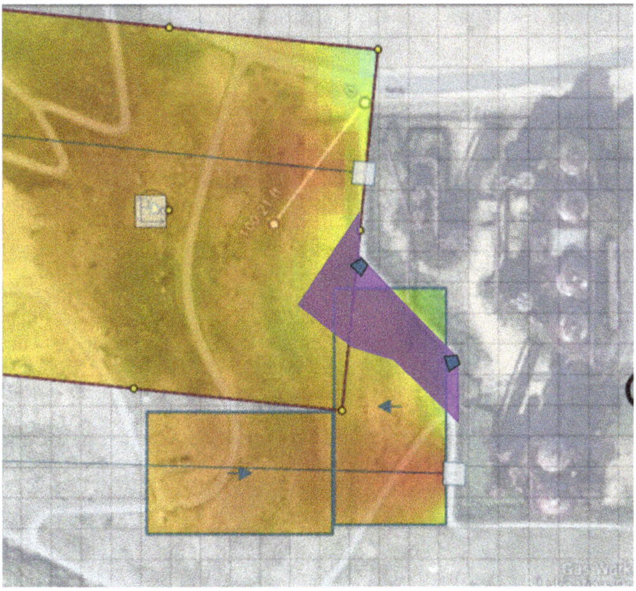

Figure 12-4

The two PA towers on the east half of the area are essentially mirror images of their western counterparts, albeit with different splay since the audience area slopes downwards instead of up, and without the need for auxiliary fills, since the first fifty feet or so is occupied by food vendors. Likewise, the south and southwest VIP areas are relatively small and easily covered by additional point source elements on tall stands operating at low SPL, which minimizes energy spill into the TV broadcast area at the SW corner of the fence line surrounding the old plant.

Once we are on site, we use our rangefinder to confirm the accuracy of the elevations we pulled from the satellite data and make any needed adjustments to the splays for each array tower.

The decentralized nature of the system requires us to give the alignment process some thought. The four principles of alignment outlined in my book can guide our order of operations here when coupled with the practical considerations: we have a front-end DSP located at the SW corner control position, which is distributing signal to each of the four PA towers, plus additional feeds for west fence line fills, south VIP and shore line VIP, plus a variety of utility feeds to the broadcast truck and the band stage on the north lawn. We also have per-box processing for each of the arrays and subwoofers, but not the networking infrastructure to control them remotely (although that's high on the list of improvements for next year's iteration). So, we will visit each tower with our laptop and measurement microphone, verify the array and subs, push our HF air compensation filters from prediction as a starting point, and then tune the array and subs to achieve the desired tonal target curve for this event. Since all four towers are the same box count with a similar amount of curvature, the EQ for each array ends up being very similar. Once we have visited all four towers, we will turn our attention to timing, and then setting level and timing for the fence line delays to fill the gaps. All of this deals with the interaction between sources, so we can work using the front-end DSP without having to reconnect our computer to the individual tower systems.

Since the east-side towers are basically twins, both firing forward into similar audience geometry, and we have no fills to deal with, we don't have any real alignment decisions to make. The west side is more complicated – the NW and SW towers are aimed on different vectors, and covering terrain of different heights with different splays – they are asymmetrical.

From up on the hill, when we get to the south edge of coverage for the NW tower, we are already hearing quite a bit of energy from the more distant SW tower. We have a rather large equal level region (what I refer to as the "seam") from very displaced sources, and so much abandon any thoughts of finding some perfect delay time to the tenth of a millisecond, as comfortable as that notion may be. No, the seam itself is large enough that there is a decent amount of "time slip" across it. We will solve this the old-fashioned way – grab a radio and fire up our favorite tuning song, calling delay values back to a colleague operating the DSP at mix position. Standing in the area where I found the time offset most disruptive, we increased delay on the nearer (NW) tower until I found it least objectionable as possible, then walked the entire transition a bit more to confirm. With source displacements this large, single-millisecond changes don't clear things up the way they might in smaller-scale situations, so I recommend bumping up or down in 5 ms or 10 ms increments until you hear it clear up, keeping in mind the goal is to do the best we can for the most listeners.

With that sorted, we go down and listen to the fence line fills from about half their coverage depth (approximately 25 feet), call for level adjustment until it matches what we were hearing in the coverage of the mains, then back up until we creep into the edge of the array coverage, and call delay times for the point sources until they snap in. Smaller seam, so timing can be more precise here. Rinse and repeat for the south shore fill, which picks up where the SW array drops out. The south VIP area fills have *no seams* with any other sub-system – they're on their own out there, with no other arrivals to hear – so we can just set them to an appropriate level and we're off to the races. Using this method, the entire alignment process took less than half an hour despite the large amount of ground to cover.

Although the approach here might seem a bit atypical to regular readers, the conceptual workflow is the same as usual – start with the large main system, set the tonality and shade for uniformity, then work your way down, bringing on the supporting system to fill the gaps and setting the timing relationships at the seams as we go.

Thirteen

Priority Shift

What makes a loudspeaker "good?" What's an example of a "good sounding" loudspeaker? Of course, if you ask ten different sound engineers, you'll get eleven different answers. At the heart of this question is a mixture of different qualities, both subjective ("I find the tonality pleasing") and objective ("This loudspeaker has very low distortion.") As discussions of this nature arise frequently amongst my colleagues (consider it an occupational hazard), I began to notice some interesting trends.

When I ask a systems engineer or system designer why they prefer certain loudspeaker products, they usually mention technical or logistical aspects — the weight, the rigging hardware, the functionality of the prediction or control software — in other words, things that have nothing to do with the *sound* of the loudspeaker, but everything to do with their day-to-day jobs and interaction with the product.

When I asked FOH engineers the same question, many of them told me they liked one brand or another because they "don't have to do much to it." After some elaboration, what many of them would tell me is that they like the "out of the box" tonality achieved by a certain system, in that they don't have to apply much EQ to the system to get it it to sound the way they want.

Loudspeaker manufacturers are undoubtedly asking end users the same questions, because the "listen to how good it sounds without EQ!" point is one I've encountered time and time again, in loudspeaker demos, tradeshow booths, and marketing materials. Clearly this is a selling point.

Now let's be clear — certainly the manufacturer's tuning that's "baked into" the loudspeaker preset should result in a loudspeaker that accomplishes what I'll call a "useful" tonal response without tinkering from the end user. Not missing a whole bunch of 250 Hz, no annoying narrow-Q peak at 2.5 kHz — that stuff should be dealt with in the manufacturer's preset. And by and large, it is. DSP is a vital ingredient in the modern era of loudspeaker design, which means manufacturers can design a loudspeaker to achieve the acoustic and mechanical performance they desire, and then use the processing preset to shape the resulting response into their desired tonality.

The red trace at the top of Figure 3-1 below shows the raw acoustic response of a single element of a popular large-format line array product measured from about 10 feet, without the manufacturer's preset in place in the amplifier. The green trace shows the response of the filtering applied by the manufacturer's preset for that loudspeaker, and the bottom pane of Figure 1 shows the resulting response in blue. (Some ripple is visible in the low-mid range due to the concrete surfaces of the warehouse in which the measurement was taken, but the point stands.)

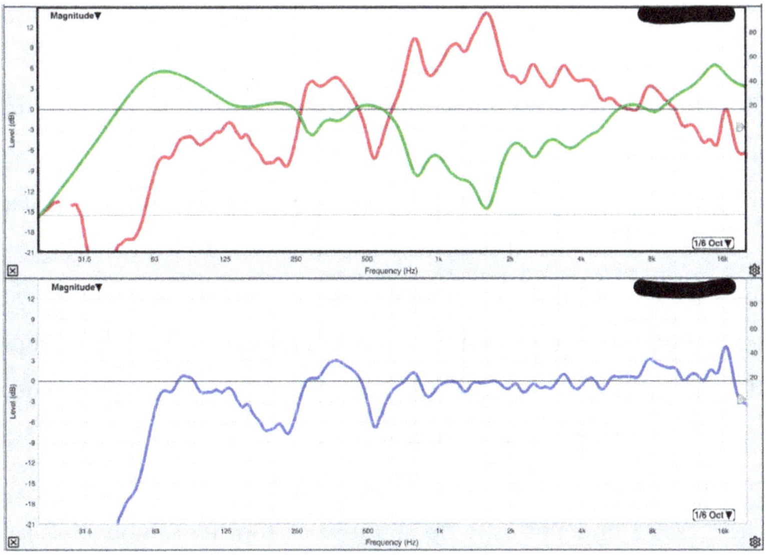

Figure 13-1

The Quest for Tonality

But besides accounting for all the high-Q peaks and dips in the acoustic response and creating a generally useful resulting response, how much effort should manufacturers be devoting to this "perfect out of the box" tonality?

I might argue that such an achievement isn't necessarily even realistically achievable — for a number of reasons. First of all — "perfect" tonality is subjective. Perfect to whom? Sure, there's a "center of the bell curve" in terms of an overall tonal contour that seems to work well for mixing live music, but many of the FOH engineers I work with regularly have specific preferences for target curve — and they're all different. So making a loudspeaker that hits a certain person's target curve perfectly by default means others will still need EQ to adjust it to their liking.

Then we have the program material. A target curve that works well for corporate, speech and musical theater might be disastrous for an EDM festival. And of course we haven't even begun to consider the list of other variables that affect the perceived tonality of a sound system — line length, box count, proximity to other loudspeakers, walls, and environmental boundaries, the reverberant characteristics of the space...

This, to me, makes the whole "let's listen with no EQ" demo quite a bit superfluous, even though multiple major manufacturers have tried to impress upon me this very point. This once crossed into absurdity when a manufacturer tried to convince me that it was "very important" that I apply no EQ to their system when aligning it for an event, to impress the show promoters and talent buyers in attendance. I tend to think a promoter would more likely be impressed by a PA that sounds excellent, rather than a PA that sounds "pretty good" plus the knowledge that no EQ was used to get it to this "pretty good" state. (None of the promoters in attendance asked to look at my processing settings.)

The reason I find all this to-do about stock tonality so non-stimulating is the simple fact that it's *really easy to change* if we don't like it. We have EQ for that. In addition to your garden-variety parametric and shelving filters, most manufacturers also include proprietary filters to change broad-strokes system tonality with a few clicks (U-Shaping, Array Morphing, CPL, etc). No one is hard-up for filters these days. It's an easy fix.

Warts and All

The elephant in the room is that many loudspeaker products *do* ship from the factory with other attributes that may not be so desirable, but unlike tonality, can't be changed by the end user, and can have a considerable effect on the sound of the system. A big category is linearity and distortion. I am very glad that in the modern era of loudspeaker design, we seem to be trending away from loudspeakers having "a sound" and towards increased linearity and high fidelity transmission. Current-generation loudspeaker products, almost without exception, have significantly lower THD (Total Harmonic Distortion) than their comparable previous-generation products. This is a good thing, and I can say that I have certainly become more sensitive to distortion products in older loudspeakers the more I listen critically to newer, more linear products.

(Some mix engineers — and indeed many listeners — enjoy the sound of certain non-linearities in musical content, such as the pleasing distortion characteristics that are hallmarks of analog tape, vinyl, tube saturators, and so forth — but those are artistic choices that belong in the mixing console, not forced upon us by a sound system struggling to maintain linearity.) The loudspeakers are part of the transmission chain, and in a typical system, contribute more distortion and non-linearity to the signal than all the other system components combined. If we don't like the distortion levels of a loudspeaker, there's nothing we can do as an end user to fix it, other than use a different loudspeaker.

Closely related to the subject of linearity is the phase response of the element. Put simply, the wider the frequency range having a flat phase response, the more of the spectrum that is leaving the loudspeaker at the same time. Remember that the magnitude and phase responses are the frequency-domain equivalent of the Impulse Response, and a flatter phase response translates to less "smearing" in the IR.

Figure 13-2 below shows the out-of-the-box responses for three large-format line array systems from common manufacturers, each 8 box-hangs with the same splay angles and same trim heights, measured from about 50 feet. Without applying any EQ, and disregarding the difference in LF extension of the purple system, we can see that they achieve extremely similar tonal responses (bottom pane).

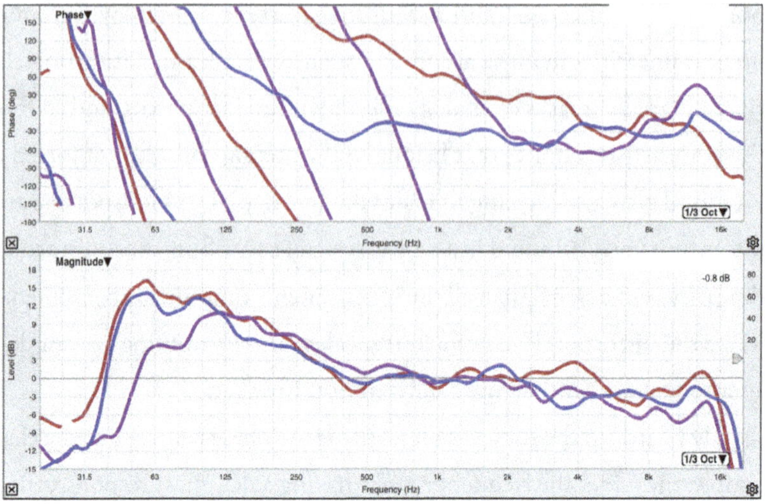

Figure 13-2

The phase responses, however, are very different. The blue system wraps around 80 Hz then stabilizes and remains phase linear all the way to the top of the spectrum. The red system wraps at 250 and then gently ramps from 180 to 0 throughout the remainder of the frequency spectrum. The purple system wraps multiple times all the way up to 2 kHz.

Despite having very similar tonalities, these three systems will sound very different for a variety of reasons, including the THD as discussed above and the polar pattern as discussed below, but in general, newer loudspeaker products are trending towards more linear phase responses and the more controlled IRs that come as a result, which I find to be a welcome trend.

Spin Me Right Round

Similarly, the polar behavior of a loudspeaker cabinet can have a profound effect on not only the listening experience of the audience, but the acoustic environment on stage. We can stand in front of a loudspeaker and adjust the EQ until we like it, but we can't control how consistent that tonality is for listeners located elsewhere in the coverage pattern — near the edges of the horn, for instance — or indeed what's coming off the back of the PA and onto the stage.

We can use AFMG's free EASE GLLViewer program to import the .GLL acoustic prediction libraries provided by most loudspeaker manufacturers and render some graphs that will allow us to examine the polar behaviors of different loudspeakers. GLLViewer can generate an impressive variety of different graphs from the raw .GLL data but for our purposes here, we will investigate how the pattern control behaves in the horizontal plane.

Imagine drawing a 20-foot wide circle in chalk on the ground and marking off the perimeter of the circle, like a clock face, in 10-degree increments. Place a loudspeaker at the center of the circle and measure its response from each location around the circumference. That's effectively what the charts below show — a transfer function overlay showing the magnitude response at a distance of 3 meters, measured every ten degrees moving horizontally around the circle to a maximum of 100° off-axis. (GLL data is often generated by placing the loudspeaker on a turntable and rotating it while the mic remains stationary, sometimes referred to as a "spin-o-rama.")

The chart is normalized on-axis, which means it uses the 0 degree (center position, on-axis) measurement as the reference, and normalizes all the traces to that, so the other traces show us *how the response changes* as we move horizontally through the coverage pattern of the loudspeaker relative to what's happening on axis.

Figure 13-3 below shows us a very well-behaved example of a dual-8" line array element. The flat red line is our on-axis 0° reference and as we sweep through the horn pattern and end up 100 degrees off axis, the HF drops more or less uniformly. We can see that the HF hits -6 dB by about 50 degrees, so we would classify this as a 100 degree horizontal dispersion.

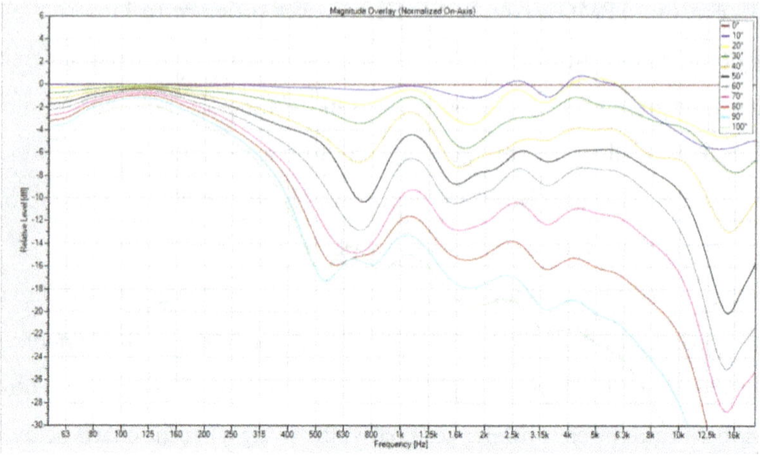

Figure 13-3

This loudspeaker exhibits an almost textbook high-frequency response, and actually behaves far better than most of its competitors. Let's look at another example, another dual-8" element, from a major loudspeaker manufacturer. (FIGURE 13-4).

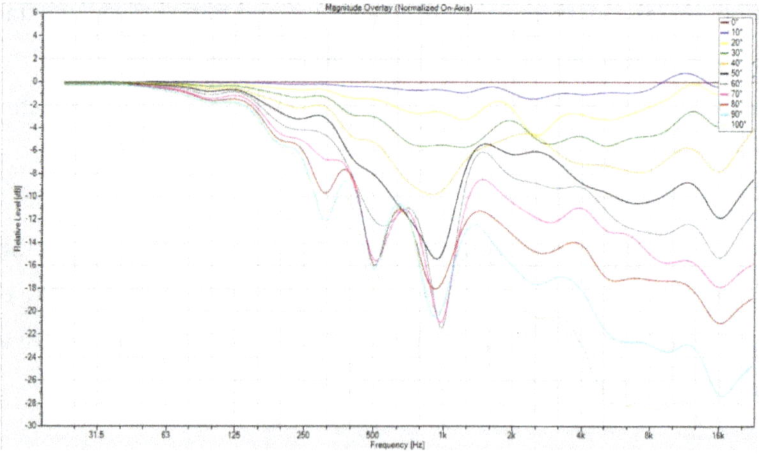

Figure 13-4

This loudspeaker has a very different tonality for the listeners near the edges of the pattern — this element is marketed as 110° dispersion, so the listening positions at the black line (+/— 50°) are fair game. The midrange plunge means that listeners in this part of the coverage hear 10 dB more 1.5 kHz than 1 kHz, and the HF falls off above that, creating a very high-mid heavy response. This translates to tonal inconsistency across its coverage area, and it *can't be fixed with EQ*. This is a design issue, not an end user adjustable parameter. Side-pattern radiation also needs to be considered whenever the array is hung next to a reflective wall, such as in a wood-surfaced orchestral hall. The tonally skewed energy reflects off the side walls and back into the listening area. Good polar behavior becomes extremely important whenever sound systems are deployed in reverberant spaces because they will affect the listening experience of all the listeners in the space, and it's something we can't fix if it's a problem. Don't like it? Need different loudspeakers.

The Focus

There's no call to action here, and that's the point — we've looked at several important ways in which loudspeaker products can be differentiated from each other, and how those differences might affect our experience as end users and listeners. None of these properties — distortion, phase response, polar pattern — are things that we have control over — they lie solely in the domain of the manufacturers' R&D teams. So why is so much of the conversation about tonality — the one thing that's trivial to change?

When auditioning sound systems, most of us tend to pay attention to the on-axis tonality by default. That's the perceptual "low hanging fruit," if you will. It's also, I would argue, one of the least important parameters by which to adjudicate a loudspeaker, simply because we can easily adjust it ourselves. Linearity and polar behavior are important considerations at the product selection phase, since we're stuck with them, so it may be a wiser strategy to select a product having the qualities that are desired for the use case, with the knowledge that we can apply EQ to achieve whatever tonality we desire.

Fourteen

Making the Map

A consistent approach to patching and labeling

Previously, I explained how I keep logical grouping of the segments of a line array covering given audience geometry, the processing driving those segments, and my measurement mic positions, such that there's a direct correlation between the measured response and the DSP output I adjust to affect that response, no guesswork. (See Shade and a Haircut for a basic example of the concept). In this article I'll explain how I extend that same logical grouping approach to the physical devices that make up a sound system (amplifiers, processors, and network switches), the patching that connects them, and the IP addresses that identify them on the control network.

Why is this important? Even if your system is 100% analog (as insisted upon by at least one of my regular FOH engineers), setting up a sound system requires connecting a lot of cables between outputs and inputs, and we want to make sure all those connections end up in the right place the first time around. This avoids unexpected results and facilitates troubleshooting — when something doesn't work, we know where to look to solve the issue. We need a system that we will always implement when patching PA drive, so we always know what our drive patch is without having to think about it too hard (I think of my naming and patching conventions the same way many mix engineers think of their channel coloring schemes. Consistency is the key to efficiency and accuracy.)

Analog Drive Patching

I try to keep my analog drive patching symmetrical, by splitting the processor outputs in half with all the PA LEFT signals starting from Output 1, and PA RIGHT signals starting from Output (N/2+1) for a processor having N outputs. For example, a processor having 16 XLR outputs — organized in two rows on the rear panel — would get all the PA LEFT outputs starting from 1 across the top row, and all the corresponding PA RIGHT outputs starting at 9 across the bottom row.

I always assign zones in the same order (Mains, Subs, Sides, 270, Front Fill, Ground Subs, Delays), skipping anything I might not be using on a particular design. (Consequently this is more or less the order in which I typically would align the system.) For example, a show consisting of L/R mains, L/R subs, front fills and outfills would be patched into a 16-output processor (such as the Galileo 616 shown below) as follows:

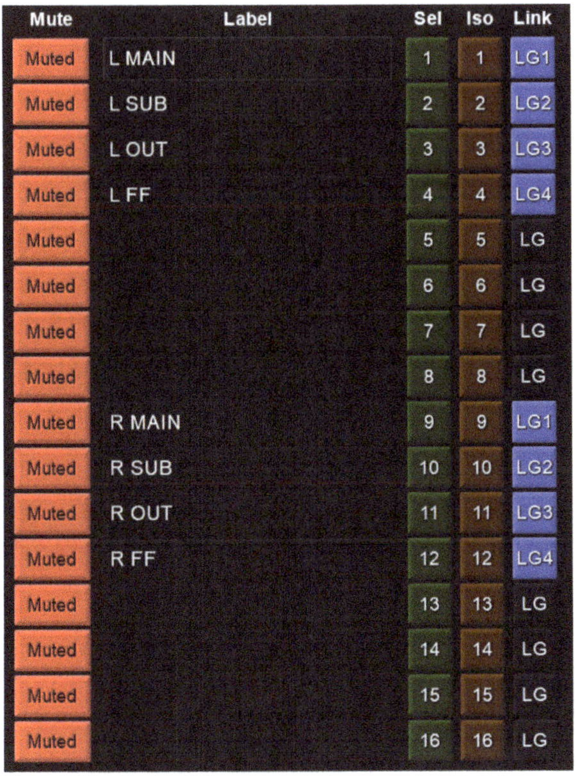

Figure 14-1

Note the logical output linking (1 and 9, 2 and 10, etc). On the rear panel of the processor, these XLR connectors will form two identical rows in the same order (MAIN, SUB, OUTFILL, FRONT FILL) which I find reduces patching errors compared to patching in pairs (Main L/R on 1/2, Sub L/R on 3/4, etc). It also allows me to easily accommodate driving subs or front fills in mono — simply eliminate outputs 10 and 12 in this example). If you're using sub-snakes, multi-core fanouts or cat5 "shuttle" adapters to run your drive lines, you'll likely find this simplifies the cabling process. It means being able to operate the unit's front-panel mute buttons in vertical pairs (first column of mutes is Mains, second column is Subs, etc), which I find handy. And it will have an additional hidden benefit when we move to networked audio distribution (Dante or Milan) as we'll see shortly.

On my AHM64 processor, which has 12 physical outputs, the same patching scheme would mean PA RIGHT starting on Output 7:

Figure 14-2

Any utility outputs such as ADA feeds would go at the end (12). And of course if you're using an 8-output processor such as an LM44, PA RIGHT would start at Output 5 under this scheme.

Adding Networked Amplifiers

Now let's look at how I incorporate the networked components (processors and amplifiers) of a typical modern PA system. Since networked amplifiers can be assigned different names depending on what file we load into the system, we need a way to definitively identify the amplifier regardless of its assigned name. When we open the software and are ready to match the physical amps in the stage racks with the virtual amps in our design, we need to take care to make those matches accurately, and whatever naming was loaded into the amps from the last show is not a foolproof method! For that reason I use Static IP addresses for each amp, laid out logically based on the amplifiers' physical locations, such that seeing the IP address in software tells me exactly where the amplifier is physically located. If you need some refreshing on the concepts of static IP address assignments and IP schemes, I highly recommend checking out *Introduction to Show Networking* by John Huntington, in which he covers both of these concepts plus a whole bunch more useful information for any audio tech working with networked gear. My networking worksheets shown in this article are a variation of the concept explained in John's book.

Unless there's a compelling reason to adopt a different scheme, I typically use IP addresses in the range of 192.168.1.xxx for all my networked devices. The last octet (three digits) is divided into three categories:

- values from 2 to 99 for devices located at FOH
- values in the 100 range for devices located Stage Left (since the snake typically runs from FOH to Stage Left, this is the first "hop" for network traffic)
- values in the 200 range for devices located Stage Right

Figure 14-3 below shows my networking worksheet for a recent event using nine RCF XPS-16K amplifiers per side.

L MAIN A	192.168.1.201	R MAIN A	192.168.1.101
L MAIN B	192.168.1.202	R MAIN B	192.168.1.102
L MAIN C	192.168.1.203	R MAIN C	192.168.1.103
L MAIN D	192.168.1.204	R MAIN D	192.168.1.104
L MAIN E	192.168.1.205	R MAIN E	192.168.1.105
L SPARE	192.168.1.206	R SPARE	192.168.1.106
L SUB OUT	192.168.1.207	R SUB OUT	192.168.1.107
L SUB MID	192.168.1.208	R SUB MID	192.168.1.108
L OUT	192.168.1.209	R OUT / FF	192.168.1.109

Figure 14-3

Notice that all the IP addresses ending in 100 are stage left (PA RIGHT) and all the addresses ending in 200 are stage right (PA LEFT). Since the amps were packaged in racks of 3, the cell outlines show the 3 racks per side. Referencing any amplifier's IP in the software tells me exactly where it's physically located. To match the amps and go online with the show file, I simply sort the discovered amps by IP address and drag and drop to the appropriate virtual amp in the file.

Note the spare amp on each side — if an amplifier were to fail, I could simply grab the spare amp (.106 or .206) and drag and drop to replace the failed amp in the project file, repatch the speaker cable and we're back up and running. And if, for example, the control software only reveals amplifiers in the .100 range, but no .200's, we know there must be an issue with the cross-stage network connection.

Incorporating Redundancy

So far we've been using the network simply for control, but as soon as we incorporate digital signal distribution as well, via a networked audio protocol such as Dante or Milan, we want our systems to be resilient against any individual device or cable failing. Although a true switch failure is relatively rare in my experience, a poorly seated cat5e cable or dusty fiber connector is much more common, especially in the live event production world, where the gear may be exposed to the elements, and handled and moved on a daily basis, and we don't want a single bad connection to stop the show.

Although there are multiple ways to skin this cat, full physical redundancy is probably the most common approach in the context of PA systems for live events. Although a full explanation of network redundancy is far beyond the scope of this article, here's the general idea: We have a network switch at FOH, which connects to the network switch in the stage left amp rack, and the network switch in the stage right amp rack, often via durable touring-grade cat5e with EtherCON barrels or fiber lines with OpticalCON connectors. This is dubbed the **Primary Network.** Now add a second network switch at FOH, and a second to the amp racks on both sides of the stage, and connect those three together in the same fashion. This is the **Secondary** network.

Our networked audio devices (consoles, processors and amplifiers) are designed to accommodate redundant network and so have two network ports that operate independently, so each device has a connection to both the Primary and Secondary networks, which both carry the same traffic. If either network experiences a failure, the other network is still intact and the performance is uninterrupted. Thus a cable can be replaced or reseated, or a network switch restarted, without stopping the show. Figure 14-4 below, a screenshot from Luminex Araneo software, shows an example of a basic redundant network topology using GigaCore network switches, a popular choice for such applications.

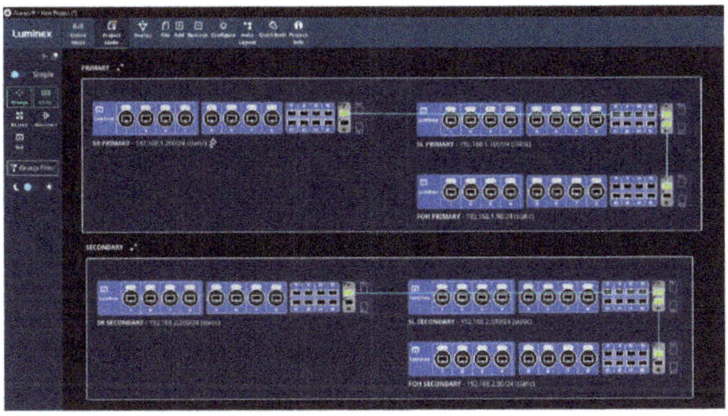

Figure 14-4

Note that there are other ways to implement various types of full or partial redundancy based on specific application requirements but for clarity we'll just stick with this one (two completely physically separate networks).

How does redundant networking fit into our IP scheme? We will change the third octet to **2** for any devices on the redundant network. Note that in the screenshot above, the last octet still obeys the rule we laid out before (2 — 99 = FOH, 100's = SL, 200's = SR) but all the switches on the redundant network have an IP address of 192.168.**2**.xxx instead of the 192.168.**1**.xxx for the primary network. Thus, the device's IP address tells us both its physical location (fourth octet) and whether it's a member of the primary or secondary network (third octet). So the first amplifier or processor on Stage Left would be assigned an IP of 192.168.1.101 for its primary network adapter, and 192.168.2.101 for its secondary network adapter, and so on.

Some network devices may not have a secondary adapter, in which case we'll just give them a 192.168.1.xxx address without a corresponding .2.

Putting it All Together

Now let's see how all these concepts come together in a real-world design example. The system I'm working with at time of writing is driven by d&b D80 amplifiers, which can accept both digital (AES3) and analog inputs. The front end processing is a pair of Meyer GALAXY processors, which send redundant Milan streams as well as analog backup. A physically redundant network (as described above) transports the Milan signals to the DS20 units on both sides of the stage, which convert the Milan signal to AES to feed the amps digitally. The analog backup from the Galaxy feeds the analog amplifier inputs directly. Thus the system has triple redundancy — if Primary Milan were to fail, Secondary Milan would keep the show going, and if for some reason both were to fail, the amps can be driven via the analog backup.

The amplifiers themselves do not have redundant control connections so they'll all be connected to the primary network only. The control computer at FOH has two network adapters installed (one for each network) so it can address and control any device on either network. The IP worksheet for the entire system looks like this (Figure 14-5):

SR			
Item	Role	ID	IP
SWITCH	SR PRIM		192.168.1.200
SWITCH	SR SEC		192.168.2.200
D80 1	MAIN 1/2	1.01	192.168.1.201
D80 2	MAIN 3/4	1.02	192.168.1.202
D80 3	MAIN 5/6	1.03	192.168.1.203
D80 4	MAIN 7/9	1.04	192.168.1.204
D80 5	MAIN 9/10	1.05	192.168.1.205
D80 6	MAIN 11/12	1.06	192.168.1.206
D80 7	MAIN 13/14	1.07	192.168.1.207
D80 8	MAIN 15/16	1.08	192.168.1.208
D80 9	SUB 2/3	2.01	192.168.1.209
D80 10	SUB 4/5	2.02	192.168.1.210
D80 11	SUB 6/7	2.03	192.168.1.211
D80 12	SUB 1	2.04	192.168.1.212
D80 13	SIDE 1/2	3.01	192.168.1.213
D80 14	SIDE 3/4	3.02	192.168.1.214
D80 15	SIDE 5/6	3.03	192.168.1.215
D80 16	SIDE 7/8	3.04	192.168.1.216
D80 17	SIDE 9/10	3.05	192.168.1.217
D80 18	SIDE 11/12	3.06	192.168.1.218
D80 19	270 1-4	4.01	192.168.1.219
D80 20	270 5-8	4.02	192.168.1.220
D80 21	GROUND	5.01	192.168.1.221
ARRAYSIGHT	L MAIN	99.01	192.168.1.241
ARRAYSIGHT	L SUB	99.03	192.168.1.242
ARRAYSIGHT	L SIDE	99.05	192.168.1.243

SL			
Item	Role	ID	IP
SWITCH	SL PRIM		192.168.1.100
SWITCH	SL SEC		192.168.2.100
D80 22	MAIN 1/2	9.01	192.168.1.101
D80 23	MAIN 3/4	9.02	192.168.1.102
D80 24	MAIN 5/6	9.03	192.168.1.103
D80 25	MAIN 7/9	9.04	192.168.1.104
D80 26	MAIN 9/10	9.05	192.168.1.105
D80 27	MAIN 11/12	9.06	192.168.1.106
D80 28	MAIN 13/14	9.07	192.168.1.107
D80 29	MAIN 15/16	9.08	192.168.1.108
D80 30	SUB 2/3	10.01	192.168.1.109
D80 31	SUB 4/5	10.02	192.168.1.110
D80 32	SUB 6/7	10.03	192.168.1.111
D80 33	SUB 1	10.04	192.168.1.112
D80 34	SIDE 1/2	11.01	192.168.1.113
D80 35	SIDE 3/4	11.02	192.168.1.114
D80 36	SIDE 5/6	11.03	192.168.1.115
D80 37	SIDE 7/8	11.04	192.168.1.116
D80 38	SIDE 9/10	11.05	192.168.1.117
D80 39	SIDE 11/12	11.06	192.168.1.118
D80 40	270 1-4	12.01	192.168.1.119
D80 41	270 5-8	12.02	192.168.1.120
D80 42	GROUND	13.01	192.168.1.121
ARRAYSIGHT	R MAIN	99.02	192.168.1.141
ARRAYSIGHT	R SUB	99.04	192.168.1.142
ARRAYSIGHT	R SIDE	99.06	192.168.1.143

FOH			
SWITCH	FOH PRIM		192.168.1.90
SWITCH	FOH SEC		192.168.2.90
GALAXY 816	FOH DSP	PRIM	192.168.1.10
		SEC	192.168.2.10
GALAXY 408	FOH DSP	PRIM	192.168.1.11
		SEC	192.168.2.11
PC		PRIM	192.168.1.99
		SEC	192.168.2.99

Figure 14-5

The colors correspond to the color coding system used on the PA carts and trunks for the tour: red for Stage Right, blue for Stage Left and Green for FOH, which helps us coordinate gear pushing to the right spot in the venue as it comes off the truck. Notice the .1. and .2. IP pairs for all the redundant devices (Galaxy processors, network switches, control computer and DS20 network bridges), and the single IP assignments for the amplifiers and inclinometers.

In addition to the network IP, the amplifiers also use a Remote ID consisting of a Subnet and a Device ID which is how the devices communicate with the R1 control software. Notice that the subnets match the front end DSP patching scheme I outlined above: PA LEFT starting on 1 for main, 2 for subs, etc, and PA RIGHT starting at 9 for main, 10 for subs, etc. These subnets match the corresponding Galaxy outputs feeding each zone. So the amp's IP and ID tell me its physical location, what part of the system it's driving, and what signal it's being fed. When it comes to digital audio transport, we reap an additional benefit of patching all the PA LEFT elements sequentially, rather than alternating between left and right: Milan signals in this system are transmitted in streams of 8 channels, so this patching scheme allows me to send one stream to PA LEFT and one stream to PA RIGHT and avoid "crossing the streams." This also reduces the likelihood of making a patching error. Figure 14-6 below is a screenshot from Milan Manager software, showing how the Milan stream containing Galaxy outputs 9-16 is driving the Stage Left DS20 for all the PA RIGHT signals.

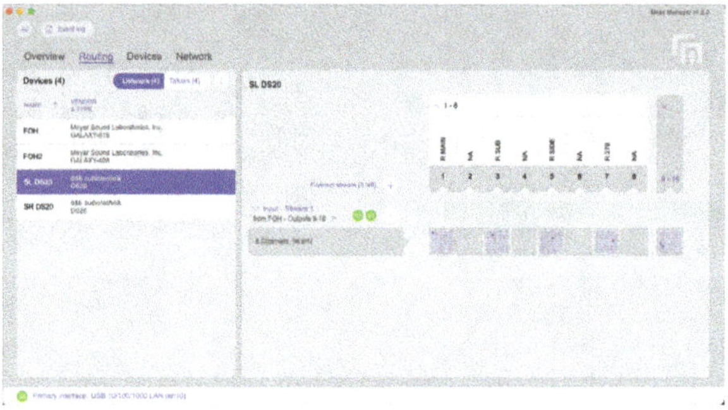

Figure 14-6

Figure 14-7

Finally, a note about security: John Huntington advises against prominent front-panel IP labels for network devices for security reasons. Although many systems engineers use a WiFi access point for roaming system adjustments, I ensure any such APs are disabled before doors open, at which point my network is air gapped and connectivity requires physical access to the network switches.

And it should go without saying — the approach detailed here is a system developed to fit my specific workflow and applications. I encourage readers working in a similar capacity to consider their own approach — and stick to it!

Fifteen

The Space Between

Investigating an oft-cited property of flown subwoofer behavior

Flying subwoofers directly behind a flown full-range array is now a well-established approach in large-scale live sound reinforcement. Some people, when learning of the technique, enquire as to whether any of the output from the subwoofers is blocked by the mains. The typical response is that the subwoofers emit low frequencies with long wavelengths (50 Hz = 22.5 ft / 6.8m) that can simply bend around the main hang and continue on its way, rather than being occluded.

I realized during a recent iteration of this explanation that I had never tested it myself, and as far as I could remember, had never seen any data in support of this explanation. Is this one of those bits of audio "common knowledge" that has been accepted as true but has a more complicated reality behind it? Is the presence of the main hang truly completely transparent to the subwoofer array? Maybe there is some sort of attenuation effect, most likely towards the upper end of the subwoofer range, but it's a good example of the Magnitude Fallacy in that it's not worth worrying about? Or maybe there's a more significant occlusion that we've been overlooking. Of course, it is scientifically sound to state that a 20 ft wavelength will not be deterred by a 3' wide line array in front of it, and our ears seem to confirm this, but it wouldn't be the first time a seemingly valid idea isn't borne out by data.

Once the question had occurred to me, the only thing to do was to test it for myself. This presents some logistical challenges as you need to be in an environment where you can lower the main hang once the PA has been flown, measure the subs unoccluded, and then raise the mains again, all without disturbing anything else on stage that might be placed beneath the PA. Also, since the main PA looms are typically draped over the top of the sub array in such a configuration, bringing just the mains back down from trim would cause a cabling crisis. A recent show afforded the opportunity to get this data, and my PA tech graciously agreed to run the mains looms alongside the sub array and independently to the cable pick, rather than over the top of the subs, to facilitate me bringing the mains in and out independently.

The test is a simple relative comparison: measure the response of the subwoofers with, and without, the mains blocking them. Then compare the traces. If the presence of the main array causes any level loss, we'll see it in the measurement.

Figure 15-1 shows the rig as deployed. The downstage rigging points for mains and flown subs are separated by 6 feet. The KSL cabinets are 24 inches deep, the subs are tipped slightly upward and the main hang has curvature so the physical space between the rear of the main hang and the front of the sub hang ends up being about 3.5 to 4 feet which is fairly typical for my designs.

Figure 15-1

I chose two mic locations out in the venue, directly on-axis with the right Main/Sub hang, approximately 200 feet (green) and 150 feet (pink) away (Figure 15-2). I also took a measurement using the mic located on venue center at mix position about 110 feet away (orange). Mix position is visible as the grey tent structure in the photos below.

Figure 15-2

Measuring the right flown sub array from these three positions — with the main hang in place in front of the sub array, but muted — yields the transfer function data shown in Figure 15-3.

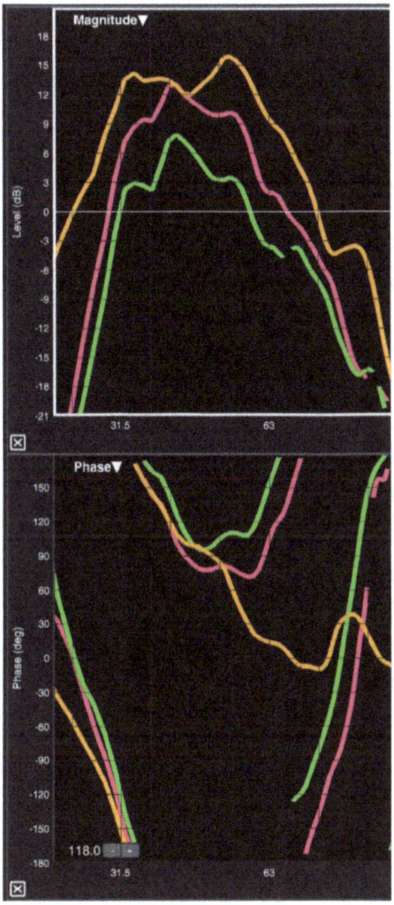

Figure 15-3

In all three cases I have added some excess measurement delay to bring the phase traces into more readable orientations, for reasons that will soon become clear. The phase deviation from the FOH trace above 63 Hz is not unexpected — remember this is a cardioid subwoofer array, so we are bound to see the effects of the pattern coming into effect as we move off axis. Such is outside of our focus here — we are interested in how these responses will (or perhaps will not) change when the main hang is lowered out of the way (Figure 15-4) and the sub array is measured again from the same locations.

Figure 15-4

I've overlaid the "with mains" and "without mains" traces as a direct comparison to see the effects of the presence of the main array in front of the subs. Here I've split the traces into three separate plots for clarity (Figure 15-5). The original (mains in place) traces retain their original colors, while the lighter colored traces in each plot are the measurements taken without the mains in the way.

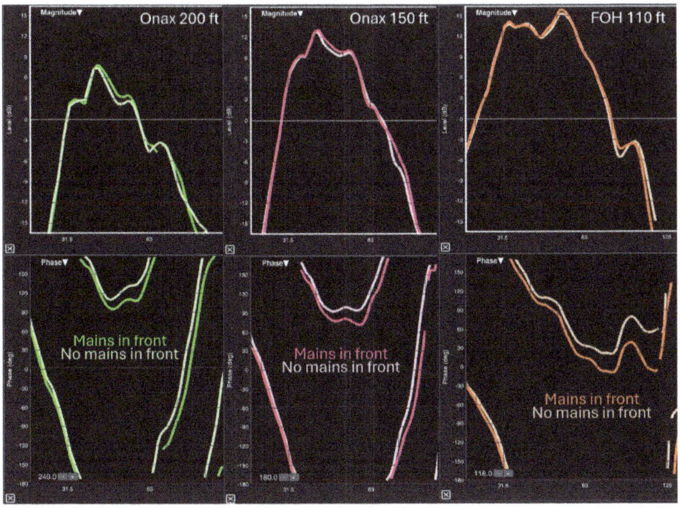

Figure 15-5

So what can we learn? The magnitude traces tell us that the conventional wisdom about the energy simply flowing around the mains hang, rather than being blocked or attenuated, seems to be accurate, as the measurement reveals no significant change in level between the two traces. Where we *do* observe a measurable change is in the phase trace. Once the obstruction of the main hang went away, the sub energy showed up *just a tad bit earlier* at all three measurement positions, as indicated by all phase traces. This makes sense — the energy takes a direct route to the listening position instead of having to take a longer path bending around the main array.

This is the same reason why using simple point-to-point distance measurement doesn't quite work for timing on cardioid sub arrays — the actually box-to-box arrival times are a bit longer due to the wraparound paths. In this case, the difference between the two scenarios was equivalent to 1.06 ms of delay, the addition of which re-aligns the sets of phase traces. Of course the exact value is of only limited academic interest and likely to be highly situation dependent, but the concept here is the valuable bit — empirical verification of the conventional wisdom that we can safely fly our subs behind our mains without loss of output.

Sixteen

Dimensional Focus: Vertical Plane

Designing, deploying, and aligning a modern sound system is a very data-driven process. This raises the question of how much data to gather. How detailed do our predictions need to be? How many measurements do we need to take? My view on this topic is that I want enough data to make a fully informed decision, but stop at the point where additional information doesn't change the decision I'll make. The recipe for efficiency is "all the necessary data, but no more." In a large venue, it might be a three or four minute walk to drop a measurement mic in certain spots in the venue, so we want to be very deliberate about where our mics are going and what we're looking to learn when we get there, to avoid unnecessary trips and minimize tuning time.

In this two-part series, we will apply the same logic to the creation of prediction models, which inherently simplify venue geometry to a series of raked planes and curved surfaces: if we only wish to spend time measuring and drawing venue architecture that will factor into our design decision, what must we include? What can we safely leave out? How can we best allocate our design time resources to the most important decisions? The answer is situation- and venue-dependent. Here we will look at two examples: one for the horizontal plane, one for the vertical.

Let's start by looking at a design for the Gorge Amphitheater in George WA.

It's easy enough to shoot the vertical profile / landing strip elevations, but modeling the horizontal curvature can be much more time consuming, and we don't want to hold up the rigging markout. Since we only need about 160° of horizontal coverage and our Mains + Sides can comfortably get us to the 180 line, our side hang will have a gentle and generous overlap with the mains while still making it to the necessary outer edge. Experience tells us that between 50° and 55° toe-out will do nicely, and the tour rigger knows how to hit that without elaboration, which allows us to completely avoid devoting any time to modeling horizontal curvature and focus our limited design time on the vertical front-to-back profiles, which will require some careful consideration.

We know where each hang is going and where it's aiming so we can build profiles quickly (FIGURE 16-1A&B). The On-Axis Profiles mode of the On-Axis app is useful for this.

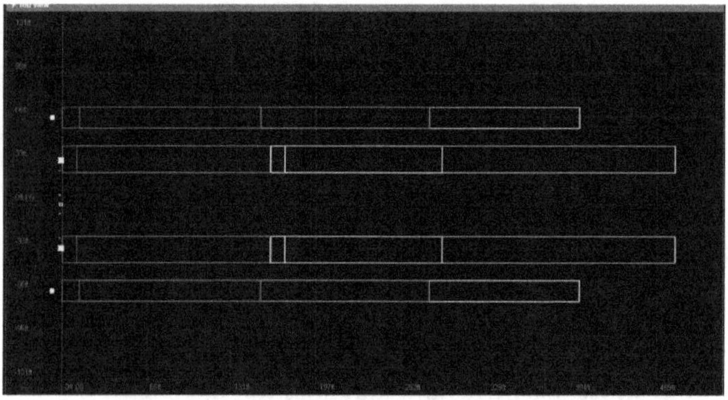

Figure 16-1A

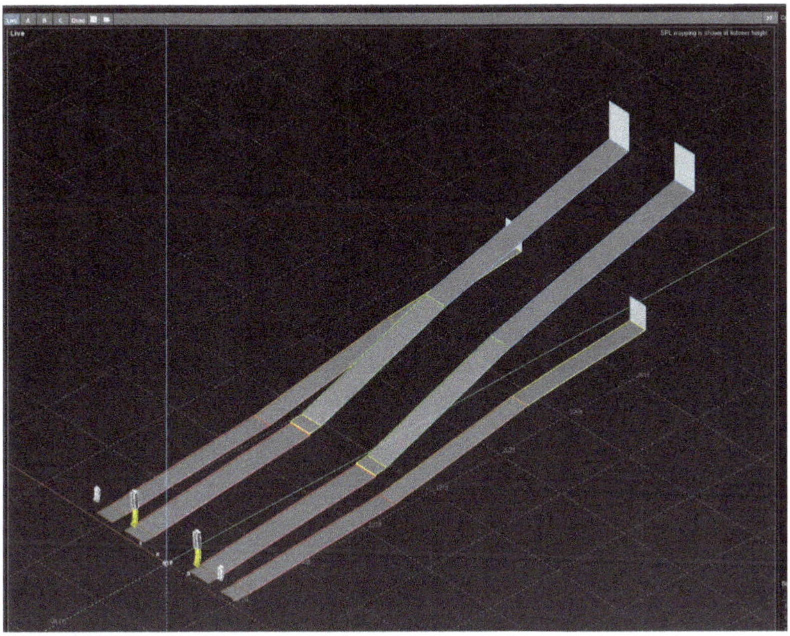

Figure 16-1B

The mains require some extra attention here. The red and green portions of the profile are handled solely by the main hangs. The blue portion of the listening plane starting at about 280 ft is the rear lawn area. The venue's lawn delay system doesn't start coverage until about 70 feet into the lawn depth, and has gaps between the towers until further back, so we need to throw some energy from the mains onto the lawn. The usual deployment on this run is 16 boxes for mains and 12 for sides, but since the mains have a much more significant throw than the sides in this particular venue, we'll steal a pair of boxes from the side hangs, and deploy the rig as 18 in the front and 10 on the side.

Although a high trim height is available to us, flying lower with a larger up-tilt means the low-mid energy lobe will naturally travel in the correct direction, meaning the ArrayProcessing algorithm won't need to work too hard to steer it, giving a more natural-sounding end result. Although the throw distance is quite significant, we are not asking the algorithm to redistribute energy in a dramatic way. (Figure 16-2)

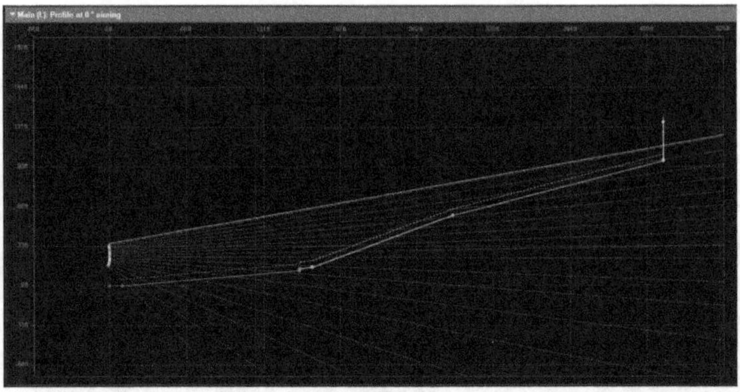

Figure 16-2

Center line impacts are more spaced on the blue plane, meaning a gentle HF pattern that feathers into the lawn system. This also avoids the "ice pick to the forehead" effect that we hear when we have a heap of 0° angles in the far field — all the low mid has dropped considerably, and the HF is lost to air absorption, leaving a speakerphone-like mid-bandpass response that's both immediately recognizable and fundamentally unpleasant. Comparing the 4 kHz and 250 Hz contours confirms that we will maintain overall tonality as the level drops with distance. (Figure 16-3)

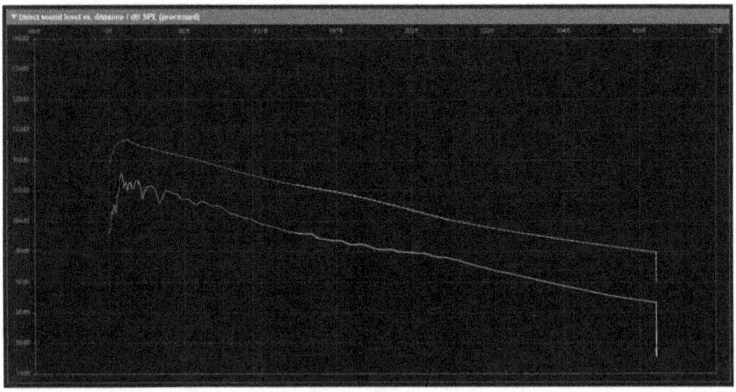

Figure 16-3

Finally, subwoofers: we're carrying flown arrays of seven per side, which is a reasonable line length for a typical arena or shed depth of around 175 — 200 feet, but not sufficient line length required to tighten the vertical pattern enough to launch the two lowest octaves all the way up the hill (see Figure 1 in Taming Red

Rocks for the importance of line length in such a situation). The "default" trim for this design is the same as the main hangs (35 feet in this case), but raising the subs to 45 feet and tipping them back 5° helps reduce the LF building between FOH and the barricade by 3 dB as well as slows the drop rate up the hill (Figure 4). Plus it looks cool (Figure 5). If we need to go further, we can apply a gentle log delay taper to steer upwards, although it didn't prove necessary in this case. And again — the LF coverage is starting to lighten up in the firsts few rows of the pit — right on schedule, as one of the design goals on this show is to keep the stage clean of LF. The sub energy is then filled back in for those listeners with four single-18" front fill subs placed in the pit, run at a much lower SPL.

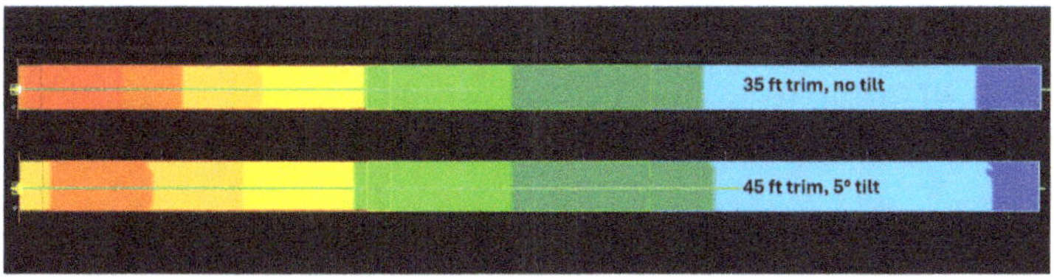

Figure 16-4

Figure 16-5

In this particular situation, adding the full 3D curvature and horizontal plane to the model doesn't change anything about how I would have chosen to deploy the system, so it's not something I wanted to devote any of my limited design time to. The OG landing-strip-style predictions can still suffice in certain situations where we don't have any surprises or complex geometry in the horizontal plane. With experience and judgement, some or all of this information can be omitted from the modeling process if time is short. (Some of my arena models infamously lack curved corners, instead built only from inclined plane for rear and side rakes. In some situations this is enough information, in others, it isn't. The skill comes in knowing which is which). If, for example, you *weren't* sure about how far out the sides should aim to accomplish the necessary coverage width, then you should model it instead of guessing, or use a low-risk mechanical deployment (single point hang can be re-angled with a rope or tie-line, or a three-point hang using a delta plate can be adjusted easily if we don't nail it the first time).

This design is an example of putting more design time and bandwidth into the vertical plane when the horizontal can be safely deprioritized. In the next chapter we'll look at an example of the opposite situation.

Seventeen

Dimensional Focus: Horizontal Plane

Previously, we looked at a real-world example of a design which benefited from extra attention paid to the vertical (front to back) plane with simple horizontal plane coverage requirements. Now we will examine the opposite situation: a venue design in which the horizontal coverage will require a special focus.

Our venue has "Auditorium" in the name, but it's a poor descriptor of its audience geometry: it's a rectangle sunken inside a larger rectangle, the whole thing equal in scale to a small arena. If you imagine draining a large swimming pool and holding a concert inside it, you've got some idea of the situation (Figure 17-1). ("Swimming pool" is also a rather apt characterization of the reverberant properties of the space, much to the chagrin of the local residents who attend shows there.)

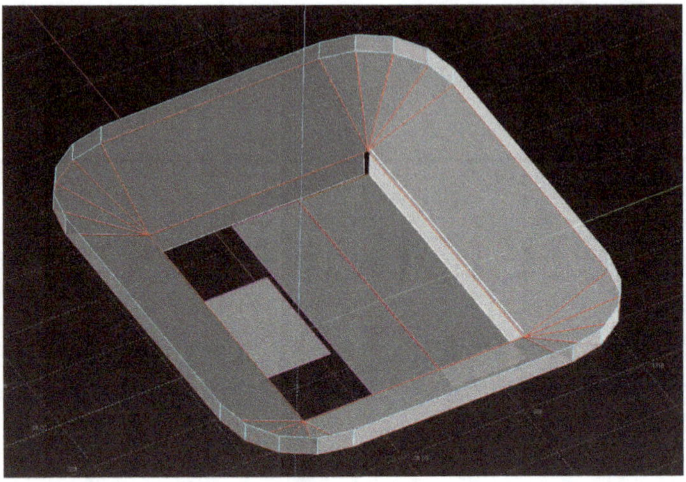

Figure 17-1

Mix position is at the rear center of the floor, immediately in front of the vertical wall. We are not scared of this wall, as it is acoustically treated, but even if it were threatening, our options for dealing with it are quite limited as we need to cover the balcony seats above, and mix position below.

These "wider than they are deep" venues tend to pose a unique challenge in terms of center gap. We have a low trim and a short throw compared to similar capacity small arenas which means we are likely to encounter a center gap, and with mix position being only about 66 feet from the stage, we need to make sure our mix engineer doesn't end up in the light coverage off the end of a horn. Figure 17-2 shows the section view and splay strategy for 12-box main hangs (truncated from 16, which would over-narrow the low-mid dispersion due to the short throw distance), nothing to see here. Our concern is in the horizontal plane.

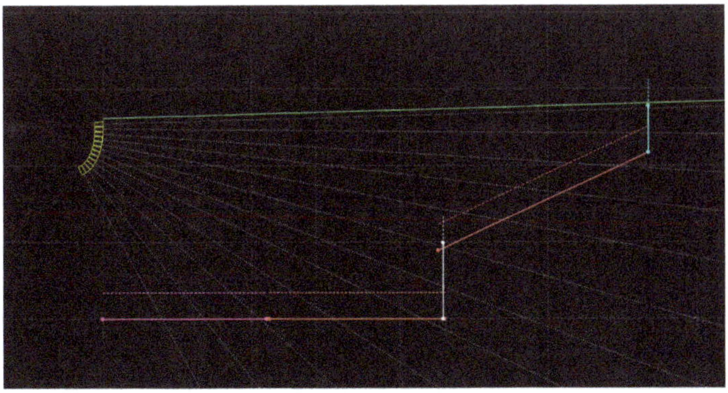

Figure 17-2

The plan view prediction shows us that our concerns were justified — coverage at mix position is light in the HF range (Figure 17-3, left). Can't have that!

Pointing the mains straight forward works great, until it doesn't. "But we always do it that way." I find convention a poor reason to continue doing something if there's a compelling reason to depart from it. A 4° toe-in fixes the issue (Figure 17-3, right).

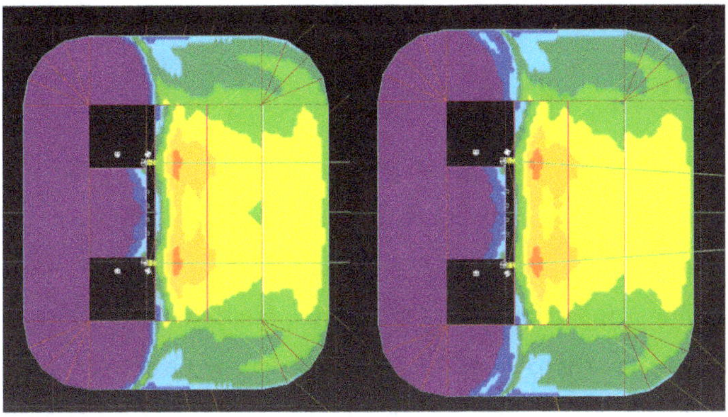

Figure 17-3

Note that due to the venue's aspect ratio, the mains don't cover around the corners and onto the side seating as they would in a typical arena, so our side hangs have more work to do (and our mains toe-in just made it worse). This model definitely needs corners :) Note that more wide boxes would have been an alternative solution here, but as we'll see now, they make themselves more useful at the top of the side hang.

Our side hang is also truncated from 12 boxes down to 8, as it doesn't have nearly as much vertical coverage to take care of it. The side will mostly concentrate its energy on the upper seating level, and only supply a tiny bit of help on the edges of the floor level, so 8 boxes is appropriate and gives us a similar splay density (impact points per foot) as the mains. The default side hang configuration is 6 wide boxes over 6 longs, and we've scratched the bottom cart of 4, so we have 6 wides over 2 longs. The wides are tremendously helpful to reach up to the rear rows of the upper seating despite the short throw. (Figure 17-4)

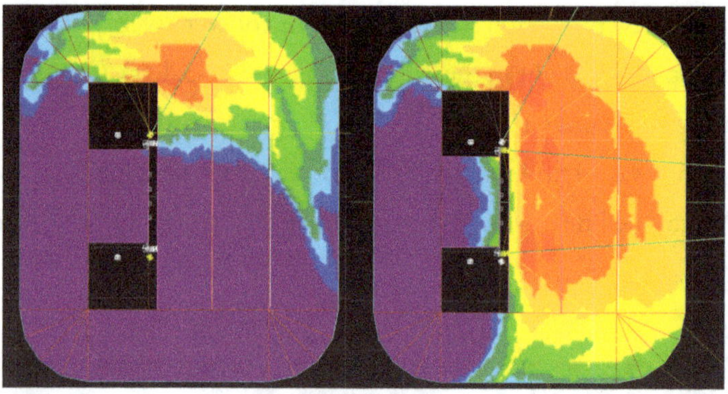

Figure 17-4

As an aside, the non-orthogonal angle of impact means we must overshoot a bit with our side hangs if we want to make sure we're covering up to the last row for the entire coverage width (see Figure 3 in Saturday In the Park for why). The simple landing strip section view of splays has limitations, and this is one of them.

Now we've covered to the 180 line, but management has asked what would be needed to open the upper seating tiers all the way to the wall. The answer: more speakers! Now we've got a 270 hang, we just need to figure out where to put it. The promoter has advised us that the upstage wall in that upper seating area is hostile, and they would only like to open those seats if we can find a way to cover them without painting the wall with excess HF.

Whenever additional radial zones or hangs are added to a sound system, the overall direct:reverberant ratio can suffer because the percentage of loudspeakers in the rig that are covering any given listening position goes down, and the percentage of loudspeakers pointed *somewhere else* goes up. Such is the nature of increased ticket sales.

To address this as responsibly as possible, we will depart from the traditional "semi-coupled" 270 hang location, hung near the mains, subs and sides, and instead slide the hang upstage, so we can hit the additional seating section head-on, which minimizes energy impacting the upstage wall as compared to a more typical hang placement. We'll also use all 80° cabinets to keep the horn pattern tight. The short throw actually helps us here. We can land the hang to keep all the needed seats in the "sweet spot" (say, within -3 dB points of the horn) and roll off outside of that. In general, a listener seated at the -6 dB point of a horn, without an adjacent source to pass the coverage baton, is experiencing a very different show than the listeners seated on axis, and I try to avoid it when possible. This is, like so many other aspects of this work, a balancing act. (Figure 17-5)

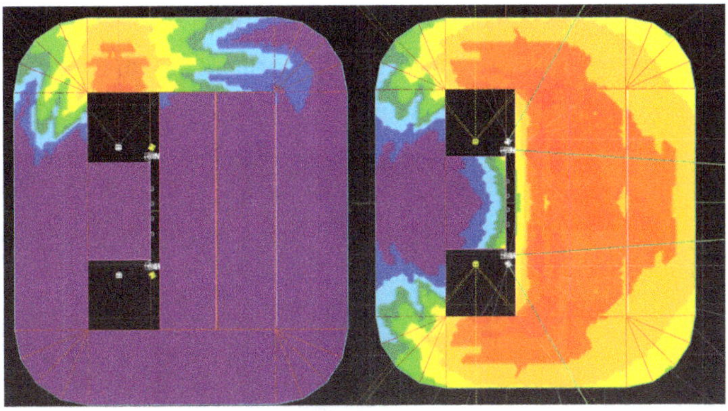

Figure 17-5

And now: the subwoofers. The default configuration for this rig (7/side flown) presents two issues. One, the line is too long and there is too much narrowing for maximum uniformity, again due to the "stout" room dimensions. Removing one and flying 6/side gives us all of the floor within the same 3 dB color bar, and most of the upper level within the next 3 dB down.

Second, and more importantly, since the main hang is now shorter, it won't curl under the subwoofer hang as usual but instead collide with it. We could space them out more, by moving the subs further upstage (not optimal, as it increases LF wash on deck) or push the mains further downstage over the audience (which requires more curvature and makes it harder to cover the first few rows, also not optimal). Solution: hang 6 subs instead of 7. We also buy ourselves a little extra wiggle room by increasing the trim height on the subs by a box.

Finally, our subs are cardioid, which actually starts working against us a bit in the 270 area. We can get those listeners a little more LF by toeing the subwoofers out. I usually avoid this technique in more standard arenas because it can put too much sub energy into the side hang area where the seats are already much closer to the arrays than the listeners in coverage of the mains (maybe the subs are -6 dB at 90° off-axis due to the cardioid pattern, but if the seats are 65 feet away instead of 180, that's a 10 dB proximity gain and only 6 dB attenuation via the cardioid pattern, leaving the sides 4 dB hot with sub energy). In this case, however, the squat aspect ratio of the room means no significant change out front and on the sides, a bit less energy on stage, and a bit more energy at the mid point of the 270 hang coverage, so we'll take it. (Figure 17-6, note the change in the yellow isobar).

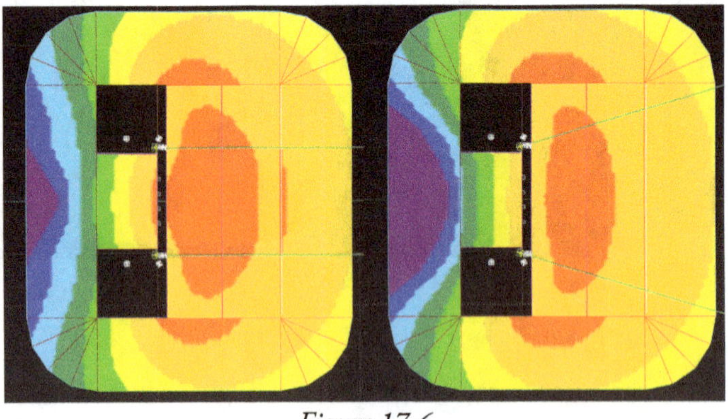

Figure 17-6

Now we have to answer the important question: can we actually hang it this way? Prediction is a wonderful frictionless, ideal environment, free of the shackles (heh) of structural steel and beam locations. The original Main, Sub and 270 locations were chosen because they can all be accomplished by simple deadhang rigging points. Luckily in this circumstance we can accomplish our off-kilter mains and subs without adding more bridles to our tour rigger's workload. Usually I seek to accomplish this via choker hitches (wrapping the beam like a P instead of an upside-down U) so the steel drops to one side or the other of the beam. By converting the mains and subs points to on-off and off-on chokers respectively, the arrays will toe out of straight alignment. (Choker hitches have more limitations than a regular deadhang basket hitch, so they are not appropriate for all situations, so this call needs to be made by the qualified rigger on site.) In this case, the points were made to the upper steel, and dropped to alternating sides of the lower steel, giving me the toe I was looking for without the capacity de-rating or beam torsion effects of a choker hitch. Luckily our tour rigger is accommodating to my design eccentricates (either that or he's a glutton for punishment) and our FOH engineer doesn't seem to mind mixing on a PA that looks like it has been involved in an earthquake. (Subs toed out. Mains toed in. Science!)

The 270 location presents an interesting wiggle when it comes to alignment: due to its "decoupling" from the main hangs and placement closer to the target listening area, it will arrive at the seam much later than the side hang. This would mean delaying the side hang significantly, which in turn means delaying the mains and subs significantly. Does it make sense to push the entire PA noticeably back in time for all listeners (including our intrepid mix engineer) to smooth a seam for a few dozen seats? This is one of the choices we must make as systems engineers. In this case I decided the answer was "no" and after listening to the rig, our FOH engineer agreed. The seam ended up being out of time by about 7 ms and neither of us found it to be overtly objectionable when walking the system, so we left it as is. Once it's aligned we are able to achieve an encouraging level of tonal consistency, which will be our safe harbor in the sea of reverberant conditions (Figure 17-7) (not all traces were taken with the subwoofers on, but no normalization was applied to them for this screenshot). The pink buildup at 60 Hz in the 270 hang was countered with EQ (it's a different model of loudspeaker than the rest of the rig, and not cardioid).

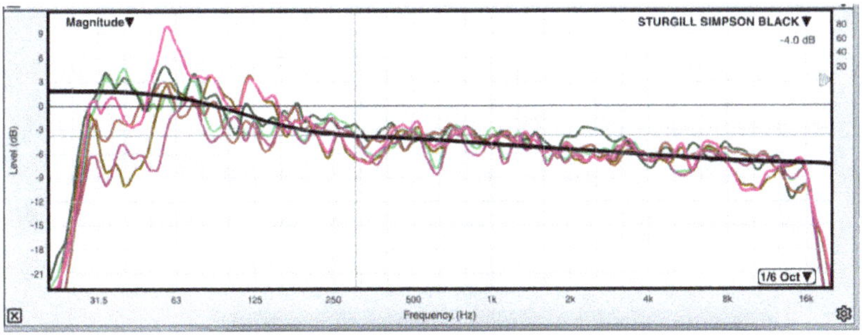

Figure 17-7

And maybe — just maybe — that might be enough to perk up the ears of locals with low expectations for the room.

Eighteen

Center Yourself

A perhaps under-discussed element of line array design and deployment is the center of gravity (COG) of the array in relationship to the flybar. Most prediction software platforms will indicate the COG to the user. If you are flying the array with a single pick point, the shackle must be directly above the COG in order for the array to have the intended vertical inclination.

Put a different way, whenever an object is suspended from a single point, it will always orient itself such that the COG of the object is directly below the pick point (See *Entertainment Rigging* by Harry Donovan for a deep exploration of this idea and its ramifications). If we pick in front of the COG, the entire array will tilt upwards. If we pick behind the COG, the entire array will tilt downwards.

In Figure 1 below, the desired vertical aiming of the array is 0°, with the flybar perfectly horizontal. The blue line indicates that the COG falls directly between two flybar pick points. If we are flying the array with a single pick, we must choose one or the other. The exact amount of error can be more depending on factors including the length and curvature of the array, weight distribution of the particular loudspeaker product, and the design of the rigging hardware.

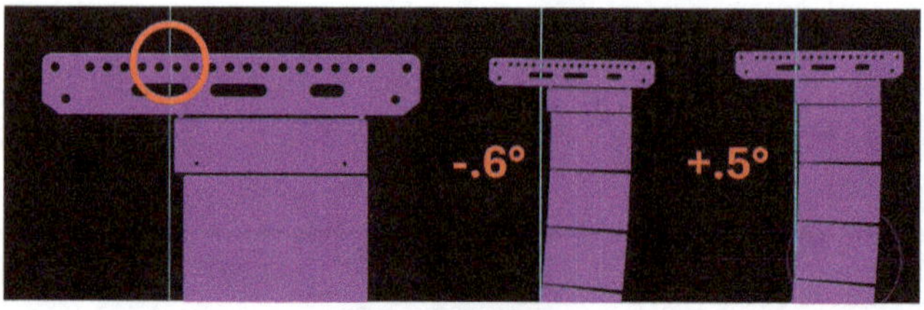

Figure 18-1

Is a half degree vertical aiming error significant? Probably not in most circumstances, but if you happen to have some 0.5° splays at the top of your array, you're now off by an entire box, which we can plainly consider to be a significant error. For the purposes of this discussion, we are ignoring other sources of error including cable weight, which can also deflect vertical inclination with a single-point pick.

Hanging an array with two points — front and back of flybar — allows us to adjust the motors independently and therefore adjust the vertical inclination once the array is in the air. Note that proper inclination can only be determined once the array is at trim height, using a laser and/or inclinometer unit mounted to the flybar.

Some people try to set proper inclination when the array is floating at deck height, using a mechanical inclinometer or smartphone app pressed against the cabinet of the bottom box. This fails for a few reasons — one of which being that most line array cabinets are trapezoidal in profile, so the cut angle of the bottom cabinet is not the aiming angle of the bottom box on-axis center line in the prediction software which is used to calculate vertical coverage.

The bigger issue, however, is that chain hoists typically don't run at the same speed, particularly when loaded unequally. If the desired inclination is set at deck height and then the array is taken to trim, the inclination has probably changed by the time the array gets to height. (Investigate this yourself by taking an inclo reading at deck height and then again at trim). With a typical two-point hang spaced by three feet, it takes less than a third of an inch of chain length difference to create the same half-degree error as above.

The Relativity of Wrong

Exactly how concerned should we be about these sub-1° inclination errors? Is it a tangible problem that we need to be vigilant about? Or is this yet another example of the Magnitude Fallacy — a real effect based in real science and math, but so small as to be negligible in real-world circumstances?

Let's see if we can put some numbers to it. Figure 18-22 shows a 17-element large-format line array hung at a trim height of 40 feet — something that we might be reasonably likely to encounter at an outdoor festival stage or amphitheater, for example. I've set the entire array to 1° splays, and aimed it so that the second box impacts the ground at 300 ft from the array.

While hanging an array of all 1° angles is probably inadvisable in terms of coverage uniformity, we can measure the ground distance between two adjacent center line impacts to see how much error in coverage is created by a 1° error in array inclination.

The red scale across the bottom shows distance from the array, and the orange rectangles span the distance between two adjacent center line impacts at milestone distances of 50, 100, 150, 200, 250 and 300 feet from the PA. The grey background grid indicates 10' increments.

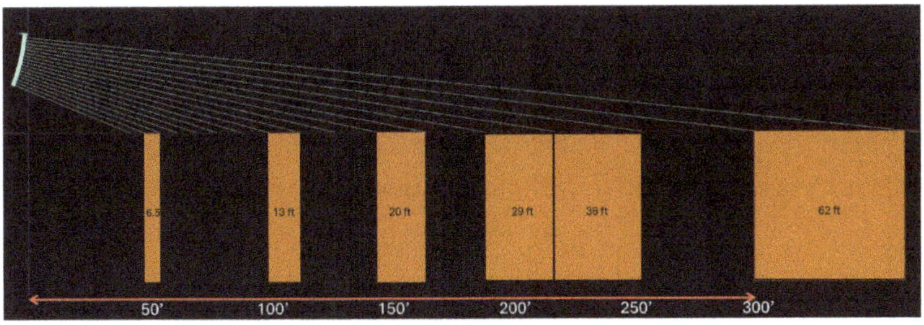

Figure 18-2

Of course, these numbers will vary greatly depending on the trim height and audience geometry but this gives us a real feel for it: up close to the PA, the error is relatively small (6.5' at a distance of 50 feet), while at a distance of 200 feet from the PA, a 1° inclination error moves our impact point by a substantial 29 feet, and we are approaching 40 linear feet of error at the 250' mark. That's about 12 rows of seats.

If our top box is supposed to cover to 300 feet, and we're 1° too high on the inclination, it's hitting at 362 feet, which is an error of 20% of the total coverage depth. Readers are encouraged to draw their own conclusions from this, but personally I wouldn't consider this negligible, especially for systems throwing long distances and impacting audience geometry at shallow angles (that means the corresponding error also gets larger with lower trim heights).

So inclinometers really are important, and this is also a good argument for building a little extra vertical splay into your design so it's a bit more robust against small aiming errors — particularly with single-point hangs, because we're unable to adjust those once the array is deployed without landing the rig and adjusting the shackle pick point, and of course the prediction software doesn't take into account the additional torque on the flybar caused by the cable loom hanging at the rear.

With a two point hang, we can of course bump our motors to accomplish the proper inclination — except when we can't. Using two motors to tilt the flybar up or down redistributes the weight between the front and rear points as the array tilts and the COG shifts forwards and backwards. At the extremes, all the weight is concentrated on one point or the other, and we're unable to tilt any further.

Note that this doesn't just apply to cases of extreme uptilt and downtilt, but also to designs with a lot of curvature: the boxes at the bottom of the "J" pull the COG substantially towards the rear, such that it's easy to end up with all the weight on the rear point even with the flybar at a flat angle.

Workarounds

So what do we do? Using an extension bar on the flybar gives us more distance between the two pick points and therefore allows a wide range of uptilt and downtilt, or more heavily curved array geometries, but in rare cases that still might not be quite enough — and sometimes we may not have an extension bar available. (In some cases we may have the ability to add a pullback rigging point for cases of extreme downtilt, but that also requires special hardware, and the number of circumstances in which we can do that is relatively low.)

If all the weight is landing at the rear point, trimming a bit lower means tilting the entire array upwards to have the top box hit at the same spot in the room, which shifts some of the weight forward, but the lower trim may not be an acceptable compromise in your particular situation. We can also play with line length. Cutting a few boxes and hanging a shorter line may bump the COG back underneath the flybar and yield a "riggable" design.

If neither of these solutions are appropriate, we can resort to a less conventional solution, which is a polite euphemism for "dirty tricks." We can game the system by using array cabinets specifically to shift the COG of the array in a way that's favorable to us by using them as a counterweight.

In the design shown on the left side of Figure 18-3, there were no extension bars available and the array frame is relatively shallow, which means we are at increased risk for COG challenges. We're in a theater with a balcony, with a lot of listeners higher than the trim height of the array, which means significant uptilt on the flybar. With all 12 boxes at the proper curvature to cover from the back of the balcony to the front row, the COG would land in front of the flybar, not possible to hang. The solution shown is to increase curvature towards the bottom of the array such that the second to last box, rather than the last box, impacts at the desired start of coverage at the front row.

On the left side of Figure 18-3, Box 12 appears to be massively over-splayed and impacting the edge of the stage, but it's not even plugged in; it's just acting as a counterweight and pulling the COG back into a usable position underneath the flybar.

Figure 18-3

That, combined with attaching the cable loom directly to the rear of the flybar to add additional weight at the rear, allowed me to accomplish the desired uptilt and send energy into the rear of the balcony. (Line length enthusiasts can feel free to connect the box to amplification and simply mute the MF/HF, using it for LF extension.)

The right side of Figure 3 shows another more subtle variation of the same principle — although the bottom box was still required for coverage into the ADA area flanked by the blue chairs, it is splayed open further than strictly required for coverage purposes, and the extra curvature was enough to nudge the COG back under the front rigging point.

I have also used this the converse of this principle — counterweight boxes at the top of the array — to circumvent a sticky rigging situation in the past. In Figure 4, we are working with a single fixed rigging point that is only 22" from the rear wall, which limits our curvature. Aiming the top box as intended and curving the array to accomplish the desired coverage would result in a collision with the wall. So we must be creative.

Figure 18-4

The top two elements create far more overshoot than I needed in the design — the design proper actually starts at box 3 — but they serve to nudge the COG forward just enough to land under the proper flybar pickpoint and achieve the necessary uptilt on the frame while avoiding collision with the rear wall and leaving a few inches behind the bottom element for cable looms to pass through, and for safely de-rigging the system after the show.

Nineteen

Hitting It Head On

In Dimensional Focus: Horizontal Plane, we looked at the potential advantages of toeing main and sub hangs in or out rather than straight ahead for the purposes of coverage uniformity. This time we'll examine the same idea from a different perspective: mix position.

Let's start by imagining a situation in which we only need mains (no sides) to cover a relatively narrow audience geometry — say, a theater — narrow enough that we can add some toe to the main hangs without creating an under-coverage situation on the sides of the room — and consider the effects of such a decision. Perhaps we have a 60 foot (18 meter) wide proscenium, which means our Mains are hung +/— 30 feet (9 meters) off center. And let's place mix position about 85 feet (26 meters) back into the audience (Figure 19-1).

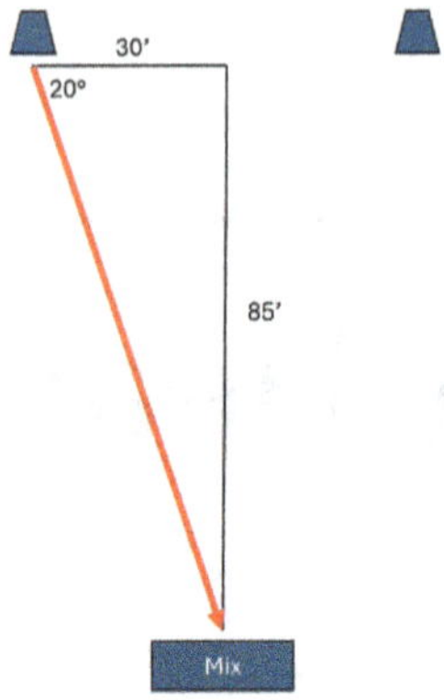

Figure 19-1

If we point the mains straight ahead in such a situation, the mix engineer will be listening to the system from about 20° off axis. We generally audition sound systems and loudspeakers on axis, right down the center of the horn, which is often where systems engineers are taught to tune systems from, and where many manufacturers make their primary voicing decisions. Nevertheless, it seems that the mix position, the place where the critical mixing decisions are made on behalf of all audience members, is more often located off axis to the sound system. This disconnect between the way loudspeakers are typically voiced and tuned and the way they're typically used in the field is a bit amusing to think about.

But surely being just 20 degrees off axis doesn't make too much of a difference? Well...

EASE GLLViewer allows us to study this "directly from the horse's mouth," by examining loudspeaker balloon data released directly by the manufacturers. We used this tool in Priority Shift to examine the polar behavior of loudspeakers in the horizontal plane, and we'll do a similar exercise again here.

The program allows us to view *normalized* magnitude data, which means we use the 0 degree on-axis magnitude response as the flat-line reference and the curve we see shows the *change* in tonality that occurs by listening from 20 degrees off axis.

Figure 19-2 below shows the 20° tonality deviation for two commonly used large-format line array cabinets, both made by the same major manufacturer. On the left, we see a cabinet that exhibits more or less "textbook" horn behavior, showing a gentle HF roll-off as we move off the center off the horn. On the right of Figure 19-2, we see a not-so-well behaved horn, as moving just 20° off center introduces some substantial deviations in the high frequency — this horn is actually brighter on the sides than it is in the middle. (And I will say that I find this particular facet of this product's behavior to be quite audible indeed.)

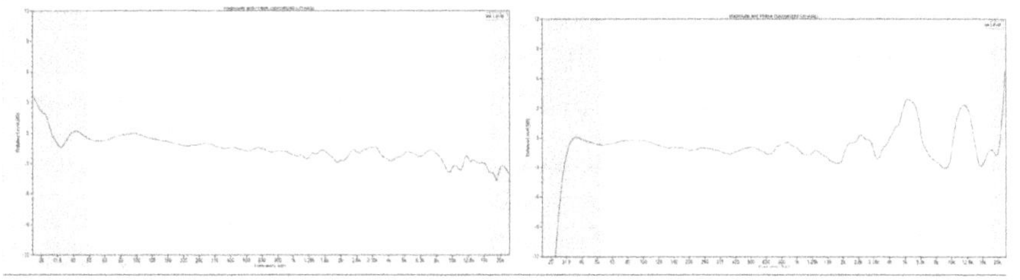

Figure 19-2

If you're thinking that's quite a substantial change for just a 20° angular shift within the loudspeaker's coverage pattern, I tend to agree with you. And this is not an isolated case. Figure 3 below shows the normalized 20° magnitude traces for three other commonly used line array cabinets from various manufacturers, all of which exhibit some substantial deviations from their on-axis response, and all in different ways. Bear in mind, this is the data as self-reported by the manufacturers, what's being used under the hood to calculate acoustic predictions.

The first thing we can glean from this is that the polar behavior of each particular loudspeaker is a major player in the severity of this effect, as the above figures illustrate, so it follows that there is substantial value for system designers to get to know the axial behavior of the loudspeakers that comprise their tool set ("Priority Shift" is a deeper dive into this concept), and to consider those behaviors when choosing between loudspeaker options for applications in which axial uniformity is a design priority. Critical listening, manufacturer data and measurement are all helpful avenues here.

Secondly, the 20 degree figure we've been using thus far wasn't arbitrary — it is a reasonable figure based on typical mid-sized venue geometries — but we must bear in mind that there is a bit of a range here as well. I pulled up model data from various venues I've worked in recently, and calculated the axial offset of mix position with respect to the main hang for each. I got a range of offsets between 15 and 30 degrees, which with some loudspeaker models is quite the difference, totally speaking. Wider displacement between mains will increase the offset at a centrally-located mix position, as will a mix position closer to the stage.

In my experience, adding a small bit of toe-in (about five degrees) to the mains seems to improve the situation at mix position markedly with some models of loudspeaker and has been well received by the mix engineers I work with regularly. Figure 4 below shows Mains hung with an 8-degree toe-in, which is about the most I typically would do for an arena configuration. In such cases, care must be taken to aim any side hangs or outfills appropriately to avoid a main-side gap, and to make sure we're still extending coverage to the edges of the required listening area. With some models of loudspeaker, the horizontal horn dispersion may not be sufficient to reach the 180-degree line with toed-in mains, and the venue aspect ratio and stage location are also ingredients in this stew, so prediction plays an important role. And of course, my solver spreadsheet (available on the Resources page of my professional website) has a handy two-point toe calculator to turn that "5 degrees" of desired to into X/Y coordinates for the riggers.

Figure 19-4

Regardless of the severity of the issue or our level of freedom to accommodate for it in the design, one thing is certain: the mix position is where our FOH engineers will be making their tonal decisions, and so we can bank on the fact that they will adjust (or ask us to adjust) the tonality of the system from that listening position until it meets their expectations. For this reason, in addition to whatever measurement and listening work we're doing on-axis to the loudspeakers, we should be carefully evaluating the resulting system tonality from mix position as well. Once the alignment is completed, I like to start my listening work at mix position, season to taste from there, and then walk the entirety of the coverage and make sure it holds up.

Twenty

Keep It Down

A man goes to the doctor and says "Doctor, it hurts when I do this!" The doctor says, "well, then don't do that!"

It's a joke that's been slung by Dads for generations, but beneath its *prima facie* silliness, there might be some wisdom lurking.

Case in point: I was tasked with designing and deploying a sound system for an event held at a casino and resort. These types of venues often feature a variety of popups and events on their grounds as entertainment for the folks staying in the hotel as well as visitors coming to the site specifically for the event. This particular event was a "pool party," in which a concert would be performed on a large mobile stage overlooking the resort's expansive pool area. By necessity, this meant that the stage — and therefore the sound system as well — would be firing directly towards the hotel's main tower, just behind the pool area.

This raises a pair of considerations: first, the sound quality for the people attending the performance and floating in the pool. Any acoustic energy from the sound system will readily bounce off the concrete and glass exterior of the hotel tower and back into the listening area, creating a highly undesirable slapback echo. As annoying as this can be for the audience members, it's even more distracting for the artists on stage, which can negatively impact the performance. In extreme cases I've also noted "second generation" slapback, when the returning echo is so severe it finds its way back into the open microphones on stage and gets amplified again. It's like a fancy delay plugin, except more terrible in every way.

Second, even though the show happened during afternoon and evening hours, it still seemed prudent to reduce the nuisance noise factor for those staying in the hotel who might be trying to rest. While this wasn't explicitly requested by the client or the vendor as a priority, it is both a first principle of sound system design (send the sound towards the audience and not everywhere else) and surely appreciated by anyone trying to sleep.

So in this case, our goals for the audience, artist, and neighbors are well aligned (which is not always the case). The best way to keep the PA from slapping off the building is, of course, not to face the concert towards the building in the first place. Beyond that, we can take steps to minimize it, but to some degree it's a fact of life in this situation.

Design Approach

What tools do we have at our disposal to accomplish this?

For starters, if we happen to be using one of the loudspeaker systems that allows for "hard avoid" beamsteering, great. There is no magic bullet here, but these systems generally do a pretty good job when programmed (and deployed!) correctly. If that tech isn't available to us, what can we do?

Maximize Trim Height

Fly the PA as high as practical (on this particular mobile stage, that means somewhere around 25 feet / 7.6m to the hook), and then angle the array more steeply downwards to minimize the amount of energy "skimming" across the ground at a low angle of incidence. Rather than the typical deck chain links between the chain hoist body and the array's flybar, I used a pear ring, providing just enough clearance for the chain bag to cleanly hang without blocking the array. This nets a bit of additional headroom on the rigging which translates into a higher achievable trim.

Less Overshoot

Usually I design with enough overshoot to fully keep the last row of seats well within the low-variance high frequency coverage. In this case, we will allow the HF to taper off just a bit towards the very rear of the pool / bar area, where the casual attendees are not highly focused on the show, which in turn gives us a bit more room for the energy to roll off before splashing against the side of the building.

Smaller Splays at Top

I usually steer clear (no pun intended) of splay angles below 1 degree, especially when the array is rigged in tension rigging (the majority case in the US), as those sub-1 degree angles can collapse to effectively zero, and in general I find that they tend not to sound very natural. The extremely small splays put a harder edge on the vertical dispersion pattern in the highest octaves, causing it to narrow before the midrange, which causes a disproportionate amount of tonal variance for those listeners on the top edge of the coverage. In this case, however, containing the high frequency output is exactly what the doctor ordered (well, right after telling us not to blast a rock concert at the side of a tall glass building with people sleeping inside). The half-degree splays at the top of the array help minimize the spill of the high frequencies past the coverage area onto the building facade, in an area where we're not highly concerned about tonal variance (folks in this area of the pool don't have much of a view of the stage anyway, so people who are deeply invested in the performance aren't watching from this location). This, combined with the reduction of overshoot (see #2 above) means a substantial reduction in high frequency splashing off the hotel glass.

Upside-down Cardioid Subwoofers

As discussed in "The Crossroads Array," I often use two-element stacks of subwoofers to allow a vertical steering component to my cardioid subwoofer deployments. We can note that when the array is deployed "upside down," with the rear-facing element on the top of the stack, we steer the rear-ward cancellation vertically, at the cost of nose-diving the forward output a bit into the ground, instead of angling up into the air. In this particular circumstance, that "shortcoming" is actually a benefit, because less energy gets sent up towards the building, and instead hugs the ground.

If you spend some time in prediction (or in the real world!) evaluating the effects of these design decisions, you will notice that none of them are truly night-and-day in terms of how much energy makes it to the hotel building behind the listening area. However, they all help a little bit — if we can get a dB or two here, and a dB or two there, the combined benefit is a substantial improvement over not giving any consideration at all to the issue. Especially in situations like this, incremental improvements are often superior to the status quo.

CHAPTER 1
INTRODUCTION

In accordance with the World Health Organization (WHO, n.d.), "Women's health is *a huge concern, and it should be prioritized under any circumstances"*. Various reports on women's health problems have attracted huge attention, such as the risk associated with childbirth and pregnancy, cervical cancer, and cardiovascular diseases. WHO reports that the women's health condition is of concern beyond reproductive and sexual factors(WHO, n.d.).Specifically, the burden of mortality and morbidity is more miserable and weaker in various countries during the reproductive age of women. As a result, it leads to humanitarian crises. The health of women during their fertile and reproductive years (ages 15 to 45) is important not only for women but also for the development of the next generation. Around the world, diseases like chronic obstructive pulmonary disease(COPD), ischemic heart disease, and stroke causes 45% of women's deaths. The cancer condition, mainly in the colon, lung, and breast, leads to a 15% death rate. Various health-related problems encountered by women in their old age are the outcomes of risk factors that arise at a young age and in their adulthood (Study, 2025) (Lu & Zhang, 2025). The various associated risk factors are sedentary lifestyles, smoking, and unhealthy diets. And this becomes hazardous for women who suffer from anaemia and malnutrition, specifically during their childbearing and pregnancy stages. The other risk factors that contribute to this include obesity, high blood pressure, and high cholesterol. Due to PCOS, the risk of sexually transmitted infections also increases, resulting in undesirable pregnancy complications like low birth-weight infants, stillbirths, congenital syphilis, and neonatal deaths. Based on some reports given by experts, cervical cancer is a common cancer identified in women worldwide, where all cases are connected virtually to genital infections (L. Yang et al., 2025). In general, disabilities and deaths among adult women are increased because of various non-communicable diseases like cancers, heart-related ailments, mental illness, injuries, and suicides. Similarly, high blood pressure is related to obesity and being overweight, and physical activities are some leading causes of risk factors among adult women (S. Wang et al., 2023). It's a disease that affects millions or trillions of women all over the world. PCOS can cause numerous complexities in the human body, like metabolic problems, infertility, and hormone abnormalities (Brady et al., 2009). Hence, obtaining an early and accurate diagnosis is challenging in clinical settings due to their complexity and heterogeneity. With the growing need for early detection of PCOS, it is crucial to integrate diverse data, such as ultrasound and omics data. Furthermore, by using this approach, deep learning models could gain a deeper understanding of PCOS etiology (M. Li et al., 2025). Recent studies have shown that combining genetic, metabolic, and clinical data can greatly enhance the precision of PCOS diagnosis, with certain models achieving an accuracy of 98.07% (Elmannai et al., 2023a). As a result, these advancements facilitate the development of personalized treatment strategies for PCOS by reducing health risks such as hypertension, diabetes, and cardiovascular disease. The evolution of diagnostic frameworks aims to improve women's health by incorporating technological advances (Dong & Rees, 2023). A new breakthrough in medical imaging has been made possible by deep learning, an advanced artificial intelligence technique. It has many promising features and is proving to be a landmark. This research aims to enhance and refine deep learning methods for PCOS detection using ultrasound images and clinical data to improve their precision, effectiveness, and interpretability. To assist clinicians in making quick, reliable diagnostic decisions, the proposed method utilizes advanced neural network architectures and data preprocessing techniques (Chakraborty et al., 2024).

1.1 OVERVIEW OF POLYCYSTIC OVARY SYNDROME (PCOS)

Polycystic ovarian syndrome, popularly known as PCOS, is the most common hormonal disorder in reproductive-aged women, characterized by higher-than-normal androgen(male) hormone levels, irregular periods, and small, fluid-filled sacs (cysts) on the ovaries (*Polycystic Ovarian Syndrome*, n.d.). Around 4-20% of the world's female population is affected by PCOS in their reproductive-age group (Ghafari et al., 2025a). Women with PCOS commonly experience high levels of the androgen hormone and irregular or prolonged menstrual periods. The follicles, which are small sacs filled with fluid that are produced by the ovaries, may be produced in vast numbers, but they may not always release eggs (Azziz et al., n.d.). The exact root cause of PCOS is still unclear. Weight reduction and early diagnosis and treatment may help reduce long-term health-related risk issues, like heart disease and type-2 diabetes. Polycystic ovary syndrome (PCOS) is the most frequent health issue affecting women (Teede et al., 2018). Endocrine disorders include PCOS as a typical reason for infertility. Infertility is the failure to successfully release the egg from the ovary in women. One of the major causes of infertility in females is polycystic ovary syndrome, where an abnormally high number of follicles develop during the ovulation phase of the menstrual cycle. A woman's ovary contains microscopic cysts called ovarian follicles, each of which contains an oocyte. Excessive body weight, increased acne and alopecia, infertility, obesity, facial hair, and irregular or non-existent periods are all signs of PCOS (Lizneva et al., 2016). Even if the precise reason for the sickness is unknown, many medical professionals believe that increased levels of androgen production in ovarian tumor cells are the root of the problem. The diagnostic criteria for this hormonal condition are hotly contested, but PCOS has been clinically validated and is often regarded as the standard used at the Rotterdam 2003 workshop. The Rotterdam criteria are based on three factors, viz., **oligo-anovulation, hyperandrogenism, and polycystic ovaries** (Fauser, 2004). PCOS is diagnosed after ruling out other related conditions that may cause similar symptoms.

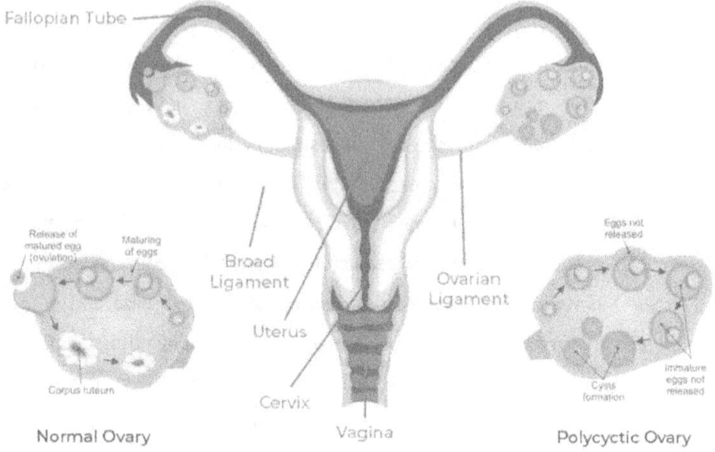

Fig. 1.1: A diagrammatic representation of Polycystic Ovary Syndrome(*PCOS*, n.d.)

Above fig. 1.1 shows the diagrammatic representation of Polycystic ovary syndrome. The ovaries contain three layers. The outer layer is called the capsule, which forms a protective layer; the middle layer is the ovarian cortex, which consists of connective tissues and contains the follicles; and the inner layer, called the medulla, contains the lymphatics and the blood vessels (Gougeon, 1996). During ovulation, one of the ovaries releases an egg (ovum). After ovulation, fertilization of the ovum by sperm results in the formation of a zygote (pregnancy). Finding PCOS at the right moment is more crucial for diagnosis and depends on the volume and number of cysts presented in the ovary. Based on the number of cysts and the size of the cysts, the ovaries in a woman's body are of three types, which are described as follows:

a) Normal Ovary

A **normal ovary** is an organ located on either side of the uterus in the female reproductive system. It's an oval, small organ that typically measures about 3 centimetres in length and resembles an almond in size and shape. The female ovary performs two primary functions namely the production of female reproductive hormones (primarily estrogen and progesterone) and the release of eggs during the menstrual cycle. A mature egg is released each month from an ovary under hormonal control known as ovulation (Vasavi & Jyothi, 2017). A normal ovary consists of an outer medulla, which hosts blood vessels, lymphatic vessels, and nerves, and an inner medulla, which contains follicles in various stage development. a normal ovary exhibits a homogenous texture and several small, fluid-filled follicles on ultrasound imaging, especially during reproductive years. The surface of the ovaries is smooth during women in their 20s and 30s,but may become inflamed due to repeated ovulations as the years go by. Hormonal and reproductive functions of the ovaries are crucial for fertility, menstrual cycle regulation, and secondary sexual traits(Qadir et al., 2023).

b) Ovarian Cyst

Ovarian cysts are sacs, usually filled with fluid, in an ovary or on its surface. Generally ovarian cysts are non-cancerous i.e. benign. It is commonly existed as a normal part of the menstrual cycle, especially during the reproductive years. Cysts like follicular cysts and corpus luteum cysts, often referred to as functional cysts, resolve on their own without treatment (Rotar et al., 2022), (Romiti et al., 2023). There are, however, other types of cysts, such as dermoid cysts, endometriomas, or cystadenomas, that may require medical attention (Mishra & Bhagat, 2025). The size of ovarian cysts varies, and they may not cause any symptoms, but generally, larger or ruptured cysts can lead to pelvic pain, bloating, irregular periods, or discomfort during intercourse (Yuktha & Nayak U, 2024). In some cases, cysts can lead to complications such as **ovarian torsion** (twisting of the ovary) or **internal bleeding**, which may require emergency care (Erdodi et al., 2025). Diagnosis is usually done through pelvic ultrasound, and treatment depends on the size, type, and symptoms. Many cysts resolve spontaneously, but persistent, symptomatic, or complex cysts may need hormonal therapy or surgical removal (Qu & Zhang, 2025). Below fig. 1.2 (Hiremath & Tegnoor, 2013) depicts all the types of ovaries in ultrasound images.

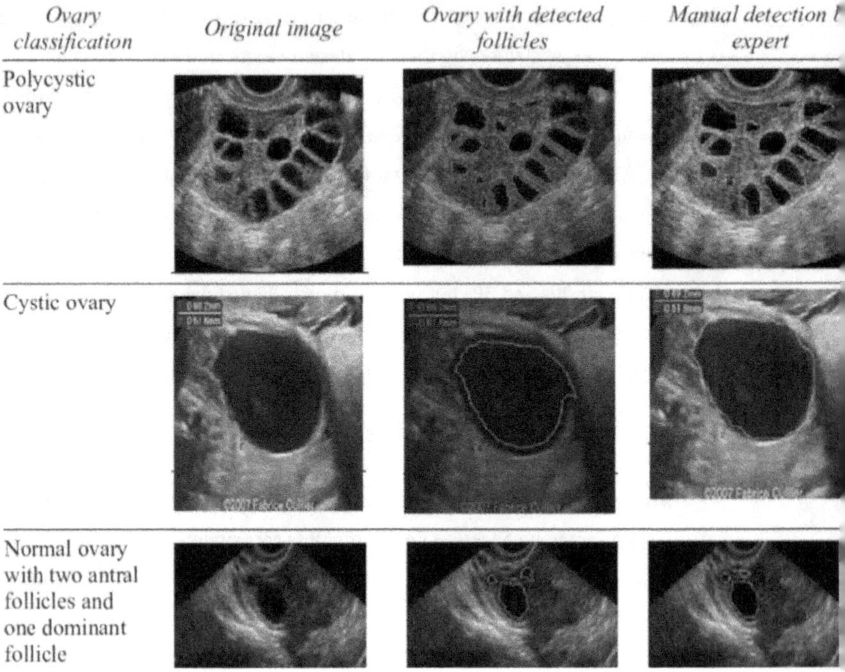

Fig. 1.2: Types of Ovaries in Ultrasound Images (Hiremath & Tegnoor, 2013)

c) Polycystic Ovary(PCO)

Polycystic ovaries contain a large number of small fluid-filled sacs of cells called follicles or cysts, typically measuring 2-9 mm in diameter (Giménez-Peralta et al., 2022), (Teede, Tay, Laven, Dokras, Tan, et al., 2023). These cysts are actually immature or arrested follicles that have not developed properly to release an egg. On ultrasound, a polycystic ovary often appears enlarged and has a characteristic string-of-pearls appearance, in which small follicles line up along the outer edge of the ovary

(Gyliene et al., 2022). This condition depicts the state of Polycystic Ovary Syndrome (PCOS), a hormonal disorder that affects women of reproductive age. Polycystic ovaries may not always show symptoms, but when accompanied by hormonal imbalance, irregularities in menstrual cycles, excessive hair growth (hirsutism), acne breakouts, and problems with ovulation, all these are symptoms of PCOS (Zad et al., 2024). But not all women with polycystic ovaries have PCOS, and some may have the ultrasound appearance without other clinical signs. The treatment of PCOS typically focuses on managing symptoms through lifestyle changes, hormonal therapy, and insulin-sensitizing medications (Cowan et al., 2023). Below, Table 1.1 (Shanmugavadivel et al., 2024) provides a summary of the types of ovaries that exist :

Table 1.1: Summary of types of ovaries

S.No.	**Type of Ovary**	**Size**	**Cyst Count**	**Image**	**Description**
1.	Normal Ovary	<10 mm	3-10 follicles	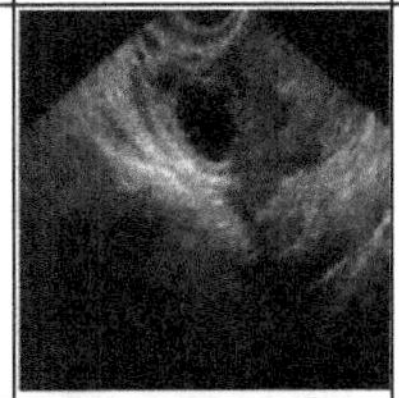	Ovaries appear to be normal with a few small follicles; no pathological changes are seen
2.	Ovarian Cyst	>20 mm (single)	Usually, solitary	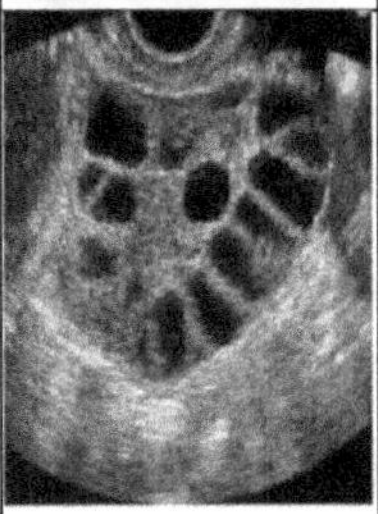	A large fluid-filled sac; may be functional or pathological

3.	Polycystic Ovary(PCO)	2-9 mm	≥12 per ovary	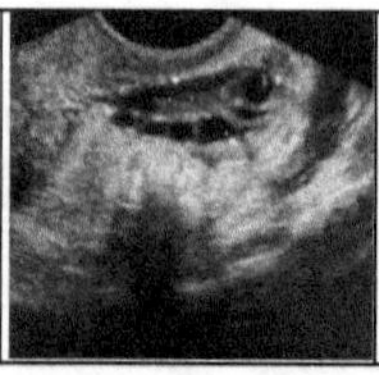	Multiple small follicles are arranged peripherally

1.2 MEDICAL SIGNS & SYMPTOMS OF POLYCYSTIC OVARY SYNDROME(PCOS)

Polycystic Ovary Syndrome (PCOS) has a variety of diverse range of medical signs and symptoms. Its presence is heterogeneous, meaning that no two women with PCOS experience identical patterns of signs and symptoms (Bai et al., 2024). In addition to complicating diagnosis and delaying clinical recognition, this variation has long-term effects on reproductive, metabolic, and psychological health. Medically, PCOS is primarily driven by three interrelated pathophysiological mechanisms: *hyperandrogenism, chronic anovulation, and insulin resistance* (Chang et al., 2024). The interaction of these mechanisms produces reproductive, dermatological, metabolic, and psychological disturbances that vary in severity depending on age, ethnicity, lifestyle, and genetic predisposition. It is important to identify the symptoms of PCOS from a medical standpoint, so that early intervention and prevention of long-term chronic diseases like infertility, type 2 diabetes, cardiovascular disease, and endometrial cancer can be accomplished as per the Rotterdam criteria (2003), oligo-anovulation, hyperandrogenism, and polycystic ovarian morphology (PCOM) are diagnostic markers of PCOS in women. The presence of either of the two symptoms is the strongest indicator of the syndrome's heterogeneity. PCOS clinical symptoms frequently progress over time, generally as menstrual irregularities and acne in adolescents, while women in their older age may experience complications such as obesity or diabetes (Helvaci & Yildiz, 2025).

1.2.1 Menstrual and Reproductive Manifestations

Menstrual dysfunction is one of the most common and significant characteristics of PCOS, where women can experience oligomenorrhea, i.e., absence of menstruation for a cyclic period longer than 35 days, or secondary amenorrhea, i.e., absence of menstruation within three consecutive months. In chronic oligoanovulation, the ovary does not release a mature oocyte in most cycles, indicating the presence of the disease (Saei Ghare Naz et al., 2020). Oligo-anovulation happens due to abnormal secretion of gonadotropin, indicated by the elevated level of luteinizing hormone (LH) to follicle-stimulating hormone (FSH) ratio, which totally disturbs the normal follicular development in the woman's body. This can also increase the risk of endometrial hyperplasia and cancer because the endometrium is continuously exposed to uncontrolled estrogen secretion (Rosenfield & Ehrmann, 2016). Reproductive symptoms encompass more than just menstrual irregularities; they also include infertility and subfertility. Based on the studies, it has been found that up to 70-80% (approx.) of women with PCOS face difficulty in conceiving due to anovulatory cycles (Melo et al., 2015). Infertility associated with PCOS primarily results from anovulation but may also be triggered by poor oocyte quality, luteal phase deficiencies, or endometrial dysfunction. Even the success rate is also low if a woman with PCOS is going through any of the assisted reproductive technologies (ART), like IVF, IUI, etc., for fertilization, implantation, and live birth rates as compared to a woman without PCOS (Cunha & Póvoa, 2021). After becoming pregnant, even premature births, gestational diabetes, preeclampsia, and early pregnancy loss are also seen in women with PCOS. In comparison with European women, south Asian women are more tend to experience severe menstrual disturbances and infertility at an earlier age. As a result of these ethnic variations, it is important to contextualize symptoms in terms of regional and genetic characteristics. With the above description it has been now concluded that symptoms of PCOS are largely associated with menstrual irregularities and reproductive issues. This health issue is not only stressful for patients, but also pass-on long-term complications related to fertility and gynecological health. Thus the clinical management of PCOS is very important and this can only be only possible by early detection of the PCOS (Kamalanathan et al., 2013).

1.2.2 Hyperandrogenic Features

Polycystic Ovary Syndrome (PCOS) is characterized by hyperandrogenism, which may be expressed clinically or biochemically. Hyperandrogenism is a condition that results from excessive ovarian and, to a lesser extent, adrenal androgen production, particularly testosterone and androstenedione. This hormonal imbalance impacts quality of life and is one of the visible manifestations of PCOS (Taieb & Feryel, 2024). The three major clinical features of hyperandrogenism include hirsutism, acne, and androgenic alopecia:

a) **Hirsutism:** It is one of the most prominent signs of hyperandrogenism, and between 60 and 70 percent of women with PCOS experience it. In hirsutism, women's hair growth in androgen-sensitive areas such as the face, chest, lower abdomen, and back is exactly the same as men's. The modified Ferriman-Gallwey (mFG) scoring system is commonly used to determine the severity of hirsutism; a score ≥8 denotes clinically severe hirsutism (Spritzer et al., 2022). This shows the importance of genetic background and hair follicle sensitivity in clinical presentation. Suffering from hirsutism may lead to psychological distress, social embarrassment, and reduced self-esteem in a woman who is going through it. Many women take medical treatment primarily due to cosmetic concerns rather than metabolic or reproductive symptoms, under the psychosocial impact of this feature (Eitah et al., 2025).

b) **Acne Breakouts:** Development of regular acne, particularly in a persistent or severe form, is another symptom of excess androgenism in PCOS. Elevated androgens stimulate sebaceous gland activity, leading to increased sebum production, follicular plugging, and itching in the skin. It is a common symptom for adolescents to experience acne in the respective age group of 20 or so, but if it persists beyond the age of 20 or is exceptionally severe, it should raise a concern regarding PCOS, which is also referred to as hyperandrogenic disorder (P. Singh et al., 2022). By age and ethnicity, acne severity ranges from 15% to 30%. Since there is no consistent correlation between acne and serum androgen levels, it is likely that sebaceous gland responsiveness as well as local androgen receptor sensitivity are important factors (Farhan et al., 2025).

c) **Androgenic Alopecia:** Androgenic alopecia causes the scalp hair to thin, especially in the vertex and frontal regions. PCOS women report experiencing this distressing symptom up to 10% of the time, despite it being less common than hirsutism. Alopecia areata varies in severity and pattern, with some women having diffuse thinning and others having localized hair loss similar to that seen in males (Fabbrocini et al., 2018). Unlike hirsutism, androgenic alopecia often shows a weak connection to the levels of androgens in the blood, highlighting how important hair follicle sensitivity is psychological distress is significant, as hair loss is firmly linked to low self-esteem, depression , and impaired quality of life . Psychological distress is significant, as hair loss is firmly linked to low self-esteem, depression, and impaired quality of life (S. Chen et al., 2022).

d) **Biochemical Hyperandrogenism:** Hyperandrogenism can also be confirmed by using parameters like measurements of testosterone, androstenedione, or dehydroepiandrosterone sulfate (DHEAS) in the body. But however, at times, most of the women with clinical hyperandrogenism may exhibit androgen levels within normal laboratory ranges, whereas some individuals with elevated androgens may not display the expected symptoms. This inconsistency may complicate the diagnosis process later (Stanczyk, 2006).Here, below Fig. 1.3 illustrates the clinical features and biomarkers of PCOS.

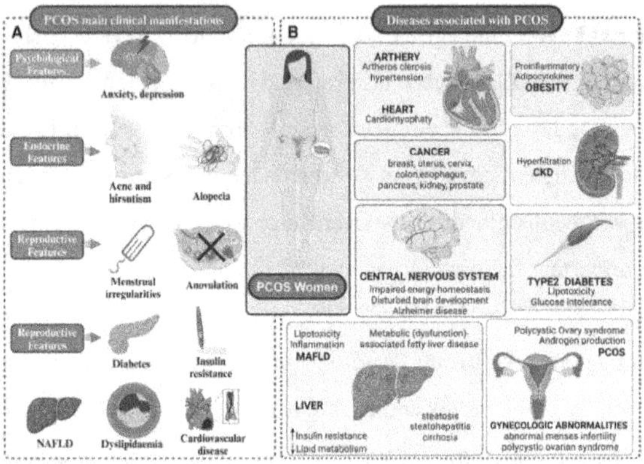

Fig. 1.3 : Clinical Markers and Indications of PCOS & a quick glance at the diseases aligned with PCOS (Y. Bin Qi et al., 2025)

1.2.3 Metabolic and Cardiovascular Manifestations:

Metabolic disturbances are also considered as one of the prominent symptoms of Polycystic Ovary Syndrome (PCOS). Apart from declining reproductive health, the metabolic disturbances also increase the risk in the long run of diseases like type-2 diabetes mellitus (T2DM) and cardiovascular disease (CVD). Conditions of insulin resistance and compensatory hyperinsulinemia are considered key pathogenic drivers of PCOS (Parua et al., 2025).

a) **Obesity and Body Fat Distribution**: Women with PCOS generally gather fat around the central part of the abdomen. Based on several studies, obesity is again highly prevalent among women with PCOS, with rates ranging from 40 to 80%.In contrast to BMI-matched controls, even lean women with PCOS had a higher proportion of visceral adiposity. By stimulating ovarian androgen production and reducing sexual hormone-binding globulin (SHBG) levels, abdominal obesity worsens hyperandrogenism (Barber et al., 2019).

b) **Insulin Resistance and Type-2 Diabetes Mellitus:** Regardless of body mass index, 50-70% of women with PCOS have insulin resistance. Hyperinsulinemia amplifies hyperandrogenism and menstrual dysfunction by suppressing hepatic SHBG synthesis and increasing the production of androgen by ovarian theca cells. Medically, insulin resistance predisposes women to impaired glucose tolerance (IGT) and progression to T2DM. Longitudinal studies have shown that women with PCOS have a five- to ten-fold higher risk of developing T2DM compared with women without PCOS (Rojas et al., 2014).

c) **Dyslipidemia:** It is another metabolic abnormality, reported in up to **70% of PCOS cases**. During this condition, lipid modifications increase triglyceride levels, reduce HDL-C(high-density lipoprotein cholesterol), and occasionally raise LDL-C(low-density lipoprotein cholesterol) this can also develops this risk of cardiovascular disease. In PCOS, dyslipidemia may occur in young women without obesity, highlighting the intrinsic metabolic risk associated with the condition(Kim & Choi, 2013).

d) **Non-Alcoholic Fatty Liver Disease (NAFLD / MASLD):** As a indicator of metabolic dysfunction of PCOS, non-alcoholic fatty liver disease (NAFLD), also popularly known as MASLD(metabolic-dysfunction-associated steatotic liver disease), is an increasingly become prevalent disease. Women suffering with PCOS disease have a 2-3-fold chances of higher risk of NAFLD as compared to age- and BMI-matched parameters. The physiological mechanism includes insulin resistance, hyperandrogenism, and central obesity, all are the indicators of accumulation of hepatic fat. In addition to being an indicator of systemic metabolic dysfunction, NAFLD/MASLD is also the cause of advanced liver diseases such as liver cirrhosis and non-alcoholic steatohepatitis (Arvanitakis et al., 2024).

e) **Cardiovascular Disease:** Atherosclerosis is commonly observed in women with PCOS, including endothelial dysfunction, elevated carotid intima-media thickness, and coronary artery calcification, but the direct association between PCOS and cardiovascular mortality remains unclear. Impacting up to 40% of women with PCOS, the aggregation of PCOS symptoms like obesity, insulin resistance, hypertension, and dyslipidemia adds to metabolic syndrome. This could definitely lead to an increase in the lifetime risk of myocardial infarction and stroke (Guan et al., 2022).

f) **Variation due to Ethnicity and Age-Related:** The metabolic symptoms of PCOS may differ among various ethnic groups. For example, in South Asian women, insulin insensitivity, obesity, and type-2 diabetes mellitus at lower BMI thresholds are more prone to happen than in European women. Age acts as an added determinant; for example, teenagers may initially display menstrual irregularities, but women in their mature age can display metabolic issues such as type-2 diabetes mellitus and dyslipidemia, which frequently become more pronounced (Chahal et al., 2020).

1.2.4 Psychological and Sleep Manifestations

Because so many symptoms arise in women due to Polycystic Ovary Syndrome (PCOS), definitely several psychosocial and sleep issues also get aggravated in the patient's body. These symptoms are not taken seriously many times, but they impact the patient's quality of life, diminish productivity, and increase healthcare utilization.

Hence, for the holistic management of PCOS it is very crucial to address all the symptoms (Dewani et al., 2023). The various psychological and sleep manifestations are described below which are as follows:-

a) **Psychological Indicators:** PCOS also affects the neuroendocrine system of the patients which develops mood swings this happens because of elevated levels of androgen hormone , insulin resistance, while symptoms like hirsutism, acne, and obesity develops psychosocial stress. All these symptoms are due to complex biological, psychological, and social problems (Fauser et al., 2012).

b) **Depression and Anxiety:** Depression and anxiety are also very common in Woman patients suffering from PCOS. In the meta-analysis of the study, it was found that around 30 to 40% of PCOS patients were depressed. PCOS women are also more likely to suffer from anxiety disorders, with an estimated occurrence of 35 to 55%. Psychosocial stressors, such as body image issues, infertility-related distress, and chronic health issues, may contribute to the pathogenesis (Cooney et al., 2017).

c) **Body-Image and Self-Esteem conditions:** Complex conditions like hirsutism, acne, and obesity can lead to a negative body image as well as low self-esteem. Adolescents and young women, particularly who face social judgment and bullying, are particularly affected by these issues. PCOS patients who feel unsatisfied with their physical appearance are more likely to suffer from depression and anxiety, illustrating the psychosocial burden of the condition (Gudiya et al., 2022).

d) **Eating Disorders:** PCOS is also characterized by aberrant eating behaviors and binge eating disorder(BED). Diabetes, obesity, and mood swings issue may trigger the inappropriate eating behaviors among PCOS patients, consequently resulting in metabolic outcomes and psychological distress (Camacho-Barcia et al., 2024).

e) **Living Standards:** A severe degraded health-related quality of life (HRQoL) is the prominent PCOS symptoms. Based on standardized assessments, including the PCOS Questionnaire (PCOSQ), women with PCOS have consistently poor reproductive health, body image issues, and mental health problems (Kite et al., 2024).

f) **Insomnia Conditions & Obstructive Sleep Apnea (OSA):** PCOS is also associated with sleep disturbances, with obstructive sleep apnea (OSA) being the most prevalent (Sam & Tasali, 2021). Obstructive sleep apnea is significantly more prevalent among women diagnosed with PCOS, regardless of their weight. In individuals with PCOS, OSA is more likely to occur between 15 and 30%, compared to approximately 5% in those without. A number of pathophysiological factors may contribute to this condition, including central obesity, insulin resistance, and elevated androgen levels, which may affect ventilatory control and upper airway stability. As a result of OSA, insulin resistance, cardiovascular risk, and mood disorders are further exacerbated, resulting in the vicious cycle of metabolic and psychological dysfunction (Doycheva & Ehrmann, 2022).

g) **Psychological and social considerations:** Ethnicity and culture appear to influence the psychological and sleep manifestations of PCOS. Compared to Caucasian women, South Asian women often experience greater psychological distress and body image dissatisfaction. In some cultures, stigmas surrounding infertility and physical appearance may exacerbate the psychological impact of PCOS, underscoring the need for culturally sensitive treatments.(Rao et al., 2022).

1.2.5 Aspects of Oncological Hazards

PCOS women also remain prone to developing various types of cancers but, generally endometrial cancer are most frequently reported, although other types of cancers like breast and ovarian cancer are also developed due to PCOS is still unclear. (Carvalho et al., 2019). Here are some details about various types of cancer diseases associated with PCOS are as follows:

a) **Endometrial Cancer:** PCOS is characterized by chronic anovulation, which exposes the endometrium to unopposed estrogen for an extended period of time without progesterone's counterbalancing effect. As a result, chronic estrogen use can increase the risk of endometrial carcinoma by causing hyperplasia of the endometrium. Endometrial cancer is two to six times more likely to develop in women with PCOS than in women without PCOS, according to epidemiological studies. Proliferation and carcinogenesis of endometrium are also influenced by obesity, insulin resistance, and metabolic syndrome (X. Li et al., 2014).

b) **Ovarian and Breast Cancer:** Some studies suggest a slightly increased risk of ovarian cancer in PCOS patients but there is no direct relationship between PCOS and Cancer is still a matter of discussion and unclear. It is possibly linked to fertility treatments or extended periods of anovulation; however, the findings are still not uniform. Inconclusive findings have been found regarding PCOS and breast cancer risk, due to the lack of substantial evidence establishing a link between PCOS and breast cancer risk (Hübinette et al., 2025). To mitigate these risks, clinical recommendations recommend assessing menstrual patterns closely, evaluating abnormal uterine bleeding promptly, and counselling on weight control and lifestyle changes. (Xue et al., 2021). Below here Table 1.2 (Gautam et al., 2025) gives the summary of the various clinical features of PCOS across different categories:

Table 1.2 : Clinical Manifestations of PCOS

S.No.	Category	Clinical Features
1.	Menstrual/Reproductive	Oligomenorrhea, amenorrhea, chronic anovulation, infertility, early pregnancy loss
2.	Hyperandrogenic	Hirsutism, acne, seborrhea, androgenic alopecia
3.	Metabolic/Cardiovascular	Central obesity, insulin insensitivity, glucose intolerance, type-2 diabetes, dyslipidemia, MASLD/NAFLD, metabolic syndrome, ↑ CVD risk
4.	Psychological	Depression, anxiety, low self-esteem, body image problems, disordered eating, and quality of life.
5.	Sleep Disorders	Obstructive sleep apnea, insomnia, poor sleep quality, excessive daytime sleepiness
6.	Oncologic Risks	Endometrial hyperplasia, ↑ risk of endometrial carcinoma, and possible ovarian cancer risk

1.2.6 Polycystic Ovary Syndrome (PCOS) : Etiology and Pathophysiology

Polycystic ovary syndrome (PCOS) is marked by various abnormalities like reproductive, metabolic, and psychological systems. Despite being so many research its exact cause is still unknown. In addition to genetic, endocrine, metabolic, and environmental factors, PCOS is recognized as a complex disorder. Despite the variety of clinical presentations among individuals and populations, PCOS is caused by overlapping and interacting processes that affect ovarian function, metabolic homeostasis, and systemic hormonal balance (Hajam et al., 2024).PCOS is etiologically and pathophysiologically associated with the following mechanisms:

a) **Excessive androgen production**
b) **Hyperinsulinemia and Insulin Insensitivity**
c) **Chronic Low-Grade inflammation**
d) **Genetic and epigenetic predisposition**
e) **Neuroendocrine dysregulation**

Each of these processes influences the degree of reproductive, metabolic, and psychological symptoms, and they all contribute differently to the PCOS phenotype (Siddiqui et al., 2022).

a) Ovarian Dysregulation and Hyperandrogenism : Excessive androgen production is one of the most common symptoms of PCOS. This is mostly caused by ovarian theca cells and has a minor contribution from the adrenal glands. Dehydroepiandrosterone sulfate (DHEAS), testosterone, and androstenedione levels were higher in women with PCOS due to increased activation of luteinizing hormone (LH) and steroidogenic hyperactivity. Polycystic ovaries are caused by this hyperandrogenic state, which disrupts folliculogenesis by stopping the maturation of ovarian follicles. As a result, many women experience infertility due to oligo-ovulation or anovulation caused by ovarian dysfunction (Bongrani et al., 2022). Additionally, PCOS is associated with adrenal hyperandrogenism, which further compounds the androgen burden.

b) Insulin Resistance and Hyperinsulinemia : Insulin insensitivity, also known as insulin resistance (IR), affects around 50-70% of women with PCOS regardless of body mass index (BMI). Insulin insensitivity in PCOS is characterized by reduced glucose absorption due to post-receptor signaling abnormalities, notably in the insulin receptor substrate (IRS)-phosphatidylinositol 3-kinase (PI3K) pathway. Pancreatic β-cells release too much insulin as a compensatory response (hyperinsulinemia) (Baptiste et al., 2010). Hyperinsulinemia exacerbates PCOS pathogenesis through multiple pathways:

i. **Androgen excess**: Ovarian theca cells are directly stimulated by insulin to produce more androgen.
ii. **Decreased SHBG**: Insulin hikes the levels of free testosterone by decreasing the hepatic production of sexual hormone-binding globulin (SHBG).
iii. **Follicular disruption**: Hyperinsulinemia alters intrafollicular signalling, impeding normal ovulation.
iv. **Metabolic consequences**: Insulin insensitivity raises the risk of type-2 diabetes mellitus (T2DM), dyslipidemia, and obesity (Ding et al., 2021).

c) **Chronic Low-Grade Inflammation:** There is growing evidence that chronic low-grade inflammation plays a part in the pathogenesis of PCOS. Inflammatory indicators such as interleukin-6 (IL-6), tumor necrosis factor-alpha (TNF-α), and C-reactive protein (CRP) are higher in women with PCOS. The increased risk of cardiovascular disease and metabolic syndrome in women with PCOS may be explained by this pro-inflammatory state.

d) **Genetic and Epigenetic Factors:** According to family and twin studies, PCOS has a considerable hereditary component; heritability estimates range between 50 and 0%. Candidate gene and genome-wide association studies (GWAS) have identified numerous loci associated with PCOS. These include genes associated with insulin signaling (INSR), androgen biosynthesis (CYP11A1, DENND1A), and gonadotropin activity (e.g., FSHR, LHCGR) (X. Li et al., 2022). In addition to basic genetic influences, epigenetic alterations such as DNA methylation, histone modifications, and microRNA regulation are increasingly recognized as contributors. Epigenetic changes may result from intrauterine exposures (e.g., maternal hyperandrogenism, gestational diabetes) or postnatal environmental influences (diet, lifestyle, endocrine-disrupting chemicals) (Y. Zhang et al., 2020). Thus, PCOS is considered a polygenic and epigenetically influenced disorder, with diverse genetic architectures across ethnic groups. PCOS is also characterized by neuroendocrine abnormalities, particularly involving the hypothalamic-pituitary axis (COMBS et al., 2021).

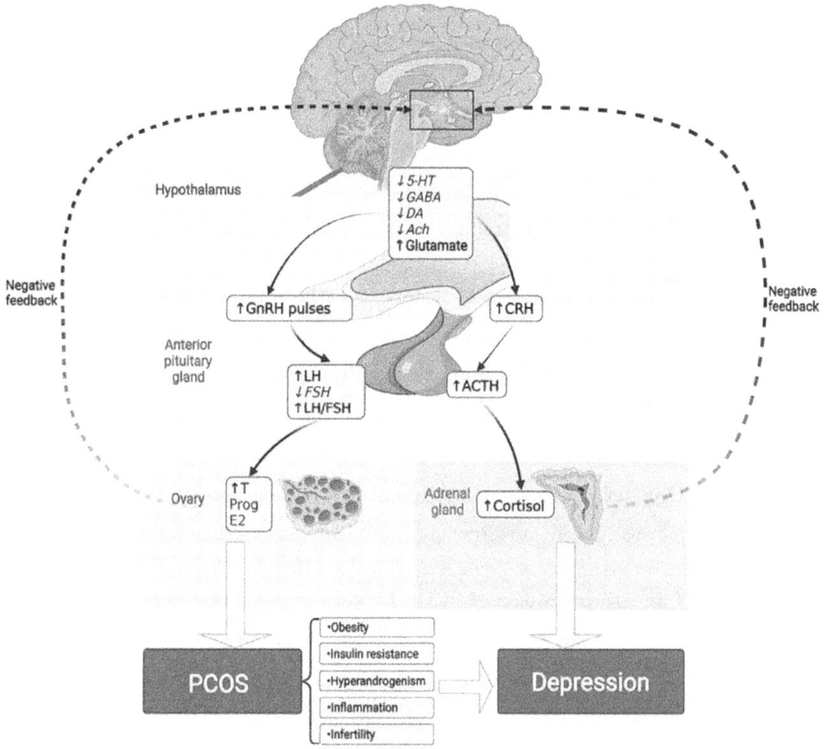

Fig. 1.4 : Relationship between insulin resistance and polycystic ovary syndrome

(Y.-B. Qi et al., 2025)

Women suffering from PCOS usually show the following symptoms:

i. The level of gonadotropin-releasing hormone (GnRH) increases in the body.
ii. Release of Luteinizing Hormone increases compared to follicle-stimulating hormone (FSH).
iii. The alteration of the LH/FSH ratio causes problems with follicle development and estrogen production.

These irregularities exacerbate ovarian androgen excess and consequently sustain anovulation. Stress and obesity may further intensify neuroendocrine dysregulation, establishing a self-perpetuating cycle (Morales-Esponda et al., 2025). Fig. 1.4 illustrates the relationship between insulin resistance and polycystic ovary syndrome.

e) Environmental and Lifestyle Influences: Although genetic and hormonal irregularities constitute the primary factors, environmental and lifestyle factors play a crucial role. Heavy-calorie diets, sedentary lifestyles, and exposure to chemicals, for example, bisphenol A, phthalates, etc., that disrupt the endocrine system may lead to worsening insulin resistance, obesity, and hormonal dysregulation in PCOS patients. Geographic and ethnic variations in PCOS phenotypes further suggest interactions between genetic predisposition and environmental exposures (Dalamaga et al., 2024).

f) Integrated Model of PCOS Pathogenesis: The pathophysiology of PCOS is most accurately characterized as a complex network of interconnected mechanisms instead of other isolated defects. PCOS patients are affected by irregularities like hyperandrogenism, insulin resistance, inflammation, and neuroendocrine disturbances (H. Liu et al., 2025). :

i. An increasing amount of insulin increases the production of androgens in the body (Ding et al., 2021).

ii. In women who produce excessive amounts of androgen, insulin sensitivity is decreased and follicular arrest occurs (Unluhizarci et al., 2021).
iii. inflammation leads to both metabolic and reproductive dysfunction (Weiss et al., 2009).
iv. The genetic susceptibility is determined by the baseline vulnerability to these perturbations (Zeber-Lubecka & Hennig, 2021).

Hence, this comprehensive description proves that every PCOS patient has different symptoms. Several mechanisms are implicated in PCOS's etiology and pathophysiology, including genetic predisposition, insulin resistance, hyperandrogenism, chronic inflammation, neuroendocrine dysregulation, etc. (Bhimwal et al., 2023). Interaction between these factors leads to clinical manifestations and exacerbates long-term health risks, including infertility, metabolic syndrome, and heart disease. In order to develop targeted therapeutic strategies, it is essential to understand these mechanisms, including lifestyle interventions and pharmacological approaches.(Mohamed et al., 2023). As shown in Table 1.3, PCOS has several mechanisms, pathophysiology features, and clinical consequences, and Fig. 1.5 illustrates PCOS patients' hormonal disorders, metabolic disorders, and inflammatory disorders.

Table 1.3: List of mechanisms, pathophysiological features, and clinical consequences of PCOS

S.No.	Mechanism	Pathophysiological Features	Clinical Consequences
1)	Hyperandrogenism (Ovarian & Adrenal)	Excessive theca cell androgen production (↑ LH sensitivity, ↑ CYP17A1 activity) (Su et al., 2025) (Heidarzadehpilehrood et al., 2022)	Hirsutism, acne, androgenic alopecia, and

			ovulatory dysfunction.
2)	Insulin Resistance & Hyperinsulinemia	Post-receptor signalling defects in the IRS-PI3K pathway (Purwar & Nagpure, 2022) (Su et al., 2025)	Insulin Resistance & Hyperinsulinemia.
3)	Low-Grade Chronic Inflammation	↑ CRP, TNF-α, IL-6 (Vasyukova et al., 2023)	Chronic Low-Grade Inflammation.
4)	Neuroendocrine Dysregulation	↑ GnRH pulse frequency (Abbara & Dhillo, 2021)	Neuroendocrine Dysregulation.

1.2.7 Diagnostic Criteria and Role of Imaging Techniques

Over the decades, the medical evaluation of polycystic ovarian syndrome (PCOS) has significantly evolved because of its varied symptoms and signs. The National Institutes of Health (NIH) established the first generally accepted PCOS criteria in 1990. They did this by defining PCOS based on two key characteristics (Joham et al., 2022):

1. **Chronic-anovulation**
2. **Clinical or biochemical hyperandrogenism**

These criteria also necessitated the elimination of other endocrine conditions, including Cushing's syndrome, hyperprolactinemia, congenital adrenal hyperplasia, and thyroid dysfunction. While the NIH definition captured the classic phenotype of PCOS, it was criticized for being too restrictive, as it excluded many women who exhibited polycystic ovarian morphology but maintained regular cycles or did not meet both features (Papadakis et al., 2019).

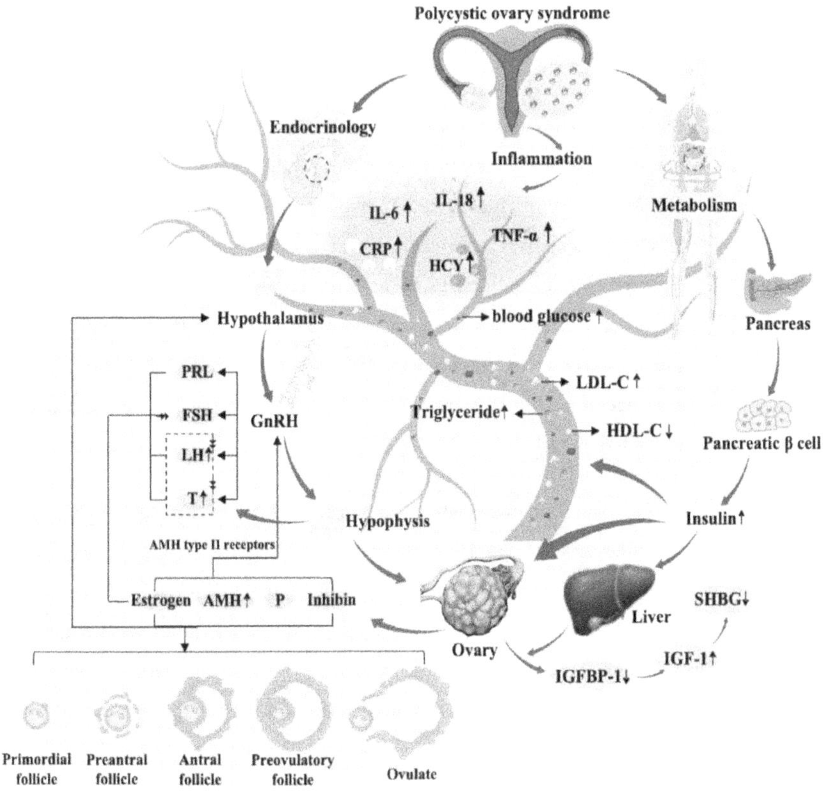

Fig. 1.5: Polycystic ovarian syndrome in hormonal disorders, the metabolism, and inflammatory disorders (Su et al., 2025)

The **Rotterdam 2003 criteria** are the most widely recognized and trusted established criteria for PCOS detection, endorsed by both the European Society of Human Reproduction and Embryology (ESHRE) and the American Society for Reproductive Medicine (ASRM) (Revised 2003 Consensus on Diagnostic Criteria and Long-Term Health Risks Related to Polycystic Ovary Syndrome (PCOS), 2004). As per these criteria, a diagnosis of PCOS is established when **any two of the following three features** are present:

1. **Oligo- or anovulation**: Defined as infrequent or absent ovulation, typically manifested as irregular menstrual cycles (more than 35 days) or amenorrhea.
2. **Clinical and/or biochemical hyperandrogenism**: Clinical signs include hirsutism, acne, and androgenic alopecia. Biochemical evidence is evaluated via elevated serum total or free testosterone levels and, in certain cases, heightened androstenedione or dehydroepiandrosterone sulfate (DHEAS) levels.
3. **Polycystic ovarian morphology (PCOM) on ultrasound**: Characterized by the presence of ≥20 follicles per ovary (2–9 mm in diameter) and an ovarian volume exceeding 10 mL.

Other diagnostic frameworks exist, emphasizing different aspects of PCOS. To declare PCOS in the patient, the **NIH 1990 criteria** approve the existence of both symptoms, viz., hyperandrogenism and ovulatory dysfunction, while the **AE-PCOS Society 2006 criteria** only focus upon hyperandrogenism along with ovarian dysfunction or PCOM. Even though there are lots of variations in symptoms and signs of PCOS, medical imaging plays a pivotal role in proving evidence of ovarian morphology (Ghafari et al., 2025b).

1.2.8 Role of Imaging in PCOS

The imaging techniques used to confirm the morphology of polycystic ovary syndrome (PCOS) play a pivotal role in its diagnosis and management. It is important to note that imaging should not be used as the sole diagnostic marker for PCOS (Di Michele et al., 2025).